I0532089

بِسْمِ اللهِ الرَّحْمَنِ الرَّحِيمِ

Praise for
Belief and the Human Mind

"*Belief and the Human Mind* offers a profound, thoughtful exploration of human belief—how faith, doubt, science, and spirituality intersect in the human psyche. "
—*Crescent Reviews*

"Dr. Bayoun writes with clarity, depth, and compassion, making complex ideas accessible while grounding them in an Islamic worldview. If you seek a book that respectfully bridges rational inquiry and spiritual insight, this work stands out as a rare and valuable contribution. "
—**Flamur Vehapi**, author of
Peace and Conflict Resolution in Islam

"Like Bayoun, I see belief as more than an intellectual assent—it is a harmony between the mind, the heart, and the lived reality of the human being. This book understands that."
—**Brandon Mayfield**, author of
Atheism Versus Belief

Revised and Updated Edition

BELIEF AND
THE HUMAN MIND

The Nature of Belief
and its Relation
to Doubt and Science

Imad M. Bayoun, Ph.D.

Foreword by
Brandon Mayfield

Crescent Books

Crescent Books

© **Imad Bayoun 2026**

To all those sincerely seeking the truth!

ACKNOWLEDGEMENTS

After God, I wish to express my deepest appreciation and indebtedness to:

My mother Nawal Daaboul and late father Mahmoud Bayoun, who made me who I am and to whom I owe everything, after God.

The greatest siblings in the world: Souha, Rania, Lama and Hani.

My late mentor Dr. Sahl Mourad, who literally opened my eyes to a new world, and whose teachings will always be central in my life.

Sh. Abdel-Jalil Mezgouri, for continuously pushing me and motivating me to complete this work

Dr. Omar Shahin, for his crucial and unconditional support throughout the work on this book.

Dr. Khaled Bahajri, Ms. Ajarat Bada, Dr. Munther El-Alami, and Dr. Nasr Alshaikh for their enriching and helpful reviews of this study.

All my wonderful friends, who are too many to list.

CONTENTS

FOREWORD

It is rare to encounter a book that speaks so directly and sincerely to the modern seeker—whether a young Muslim wrestling with inherited faith, a convert newly navigating its depth, or anyone caught between the poles of conviction and confusion. Dr. Imad Bayoun's Belief and the Human Mind is such a book: courageous, intellectually grounded, and spiritually generous.

What immediately sets this work apart is its honesty. Dr. Bayoun opens not with dogma, but with reflection—acknowledging the ebb and flow of his own belief and inviting readers to do the same. He does not present faith as a fixed state, but as a journey—one marked by questions, strengthened by clarity, and matured through lived experience. Doubt is not dismissed, but examined, clarified, and even embraced as part of the path to deeper certainty.

And yet, this is not merely a book about doubt—it is also a book about certainty. Bayoun gives careful attention to the ways in which belief becomes grounded—not through emotional impulse alone, but through reason, reflection, and experiential understanding. He skillfully introduces the Qur'anic and classical framework of yaqeen (certainty), exploring how one moves from knowledge of something, to direct perception, and finally to lived truth. His analogy of the ocean—hearing of it, seeing it, then swimming in it—beautifully illustrates how belief evolves from theory to reality.

A central contribution of this book is its nuanced treatment of science and revelation. Dr. Bayoun does not set them at odds, but he is clear-eyed about their differences. He reminds us that while the Qur'an claims to be a complete guide for human purpose, science makes no such claim. Scientific knowledge, despite its remarkable advances, remains incomplete—limited to the measurable and constantly evolving. It may answer the how, but not the why. Religion, on the other hand, offers purpose, direction, and meaning—truths that science cannot reach because they are rooted in the unseen and the metaphysical.

Dr. Bayoun also acknowledges that both science and revelation carry degrees of certainty and ambiguity. Not every scientific conclusion is definitive, just as not every religious narration is conclusive or unambiguous. Some verses in the Qur'an, as the text itself declares, are allegorical or open to interpretation. A mature faith, Bayoun argues, recognizes these distinctions and resists the temptation to bend either science or scripture to fit a desired narrative.

As someone who has wrestled with many of these same themes in my own writing—particularly in *Atheism Versus Belief*—I found *Belief and the Human Mind* both affirming and enriching. Like Dr. Bayoun, I see belief as more than an intellectual assent—it is a harmony between the mind, the heart, and the lived reality of the human being. This book understands that. It refuses to flatten belief into certainty or to inflate doubt into paralysis. It allows space to question, but also insists on the legitimacy of arriving at truth.

Dr. Bayoun's voice is a necessary one—gentle yet firm, principled yet compassionate. In a time when many religious spaces dismiss honest inquiry and many secular spaces dismiss faith as irrational, *Belief and the Human Mind* charts a middle path. It does not offer final answers to every question, but it shows that we already have enough to move forward. And that, perhaps, is the essence of sincere belief.

Brandon Mayfield
Author of *Atheism Versus Belief*

In the Name of God, the Beneficent, the All-Merciful

PREFACE

For many years, I had been lecturing about the importance of belief, what it should be like, and its role in people's lives. I thought I was good, myself. After all, I've been teaching others on how to strengthen their belief; I surely must have it myself!

I was busy researching, teaching, mentoring, and over-glorifying my own experience... then reality hit! Faced with some major challenges, one after the other, faith quickly started eroding away. I found out my faith wasn't as strong as I thought, certainly not as I was projecting.

Hardships certainly expose our truth; they show us our own reality. I was forced to investigate further and look deeper. More than any other reason, I was looking to answer some of my own questions. I did not want to learn in order to teach; I had been doing that for the longest time. I wanted to learn in order to find my own belief, for myself. And, anyway, if I have it, I can certainly teach it more efficiently!

I had an urgent need to understand what my own belief is supposed to be like, what it comprises, and ways to strengthen it. I am hoping I got it right this time!

13

INTRODUCTION

We are continuously reminded by our own reality of the significance of belief in our life. Our lives are mostly reflections of our set of beliefs. It is the perspective through which we see ourselves and the world. And whether we realize it or not, we mostly build our life goals and set our priorities in light of our beliefs. These beliefs become a major driving force and source of motivation for us.

Additionally, we all have a natural tendency to look for the truth. Curiosity and the affinity to understand have always been major driving forces. Throughout history, people's urge to learn has propelled them into conducting extensive research and taking remote exploration trips. We want to know, to understand. As Christopher McCandless[1] who lost his life "into the wild" strongly said, "Rather than love, than money, than faith, than fame, than fairness, give me truth."[1] I do believe many people have the same affinity, even if not as strongly. Man is eager to know the purpose of his existence, to understand *why* things are the way they are. It gives him the needed direction and the necessary grounding force, while dealing with his own desires and insecurities and in meeting various environmental pressures.

---

[1] Christopher Johnson McCandless (1968–1992) was an American adventurer who set out into the Alaskan wilderness in pursuit of a period of solitary and simple living. Nearly four months later, his lifeless body was discovered in an old bus he had turned into a makeshift shelter.

No one would doubt the value of the belief in God in people's lives. Even those that do not believe in God still believe in the value of such belief. Al-Qaradawi, an Islamic scholar, once said, "don't spread doubts about the Existence of God, for if it wasn't for that belief, my wife would betray me and my servant would steal from me."[2,2] The Russian philosopher Dostoyevsky[3] famously wrote, "if God does not exist, then everything is permissible."[3] Al-Jisr[4], a Lebanese theologian, contends that belief in God is the foundation of good morals, the source of patience during hardships, the foundation of contentment and happiness, and the strong knot between humanity and its noble ideals.

One might further contend that people have a natural spiritual inclination, always longing to believe in something supreme. This propensity has always been central in people's pursuit of happiness. People seem to always default toward religion and spirituality, regardless of cultures and geographical locations. In a demographic study based on analyzing more than 2,500 censuses, surveys, and population registers, The Pew Research Center found that 84% of the world population is affiliated with some religion.[4] Only 16% (1.1 billion) have no religious affiliation.[5] But, interestingly, even among these, many still hold some religious or spiritual beliefs (such as belief in some divine or universal spirit), despite not identifying with a particular faith.

In their quest to find the truth and the Divine, people have searched and arrived to proving God/belief in their own ways:
- Some people have simply answered the call of human nature (Fitrah), as stated in this verse:

---

[2] Yusuf Al-Qaradawi (1926 – 2022 ;يوسف القرضاوي) was an Egyptian Islamic scholar, and chairman of the International Union of Muslim Scholars.
[3] Fyodor Mikhailovich Dostoyevsky (1821 – 1881) was a Russian novelist, short story writer, essayist, journalist and philosopher.
[4] Nadim Al-Jisr (1897 – 1980) was a Lebanese theologian, politician, and author.

قَالَتْ رُسُلُهُمْ أَفِي اللَّهِ شَكٌّ فَاطِرِ السَّمَاوَاتِ وَالْأَرْضِ

Their apostles said: "Is there a doubt about God, the Creator of the heavens and the earth?"[6]

- Other people have arrived at the same conclusion through the laws of cause and effect. A perfectly created universe must have a Creator. For everything existent, there is someone that brought it into existence; for every effect there is a cause. It is a basic logic that allows humans to answer many questions in life.

- Still, other people have come to the same conclusion by considering the benefit of such belief. If believing in something brings benefit in case it were true and no loss if it were false, then the smart choice would be to believe. They follow a pragmatic approach, judging beliefs and ideas by their impacts. Something is accepted if it results in a benefit (which itself is determined subjectively), and vice versa. The Qur'an does recognize the importance of that logic in people's lives as in the following verses:

مَنْ كَانَ يُرِيدُ ثَوَابَ الدُّنْيَا فَعِنْدَ اللَّهِ ثَوَابُ الدُّنْيَا وَالْآخِرَةِ وَكَانَ اللَّهُ سَمِيعًا بَصِيرًا

If any one desires a reward in this life, in God's (gift) is the reward (both) of this life and of the Hereafter: for God is He that heareth and seeth (all things).[7]

وَقِيلَ لِلَّذِينَ اتَّقَوْا مَاذَا أَنْزَلَ رَبُّكُمْ قَالُوا خَيْرًا لِلَّذِينَ أَحْسَنُوا فِي هَذِهِ الدُّنْيَا حَسَنَةٌ وَلَدَارُ الْآخِرَةِ خَيْرٌ وَلَنِعْمَ دَارُ الْمُتَّقِينَ

To the righteous (when) it is said, "What is it that your Lord has revealed?" they say, "All that is good." To those who do good, there is good in this world, and the Home of the Hereafter is even better and excellent indeed is the Home of the righteous.[8]

Thus, the pragmatic approach might be legitimate, but some important considerations need to be taken. Firstly, we believe the truth deserves regard for itself (for being the truth), whether it brings benefits or not. But we believe that the truth can only be beneficial, anyway. Secondly, our definition of what is beneficial may be more comprehensive and different from the way others define it. We don't just look at the physical and individual considerations, but beyond that, we consider metaphysical, spiritual, intellectual, and communal benefits, amongst others.

One such major metaphysical benefit is the sense of continuity, of eternity. Believing in God and an Afterlife expands one's existence to include another life after the present one is finished. This was summed up by the famous statement of the Prophet's ﷺ companion Rab`ee Bin `Amer to Rustom before the beginning of the battle of Al-Qadisiyyah:

الله ابتعثنا لنخرج من شاء من عبادة العباد إلى عبادة الله، ومن ضيق الدنيا إلى سعتها، ومن جور الأديان إلى عدل الإسلام...

*Allah sent us to lead people out of worshiping one another into worshiping Allah, and **from the tightness of this world to its vastness**, and from the injustice of other religions into the justice of Islam...*[9]

A belief in the Hereafter certainly brings a sense of vastness, as the person does not consider death to be the end of his existence, but rather a gate into another everlasting life.

Such belief evidently carries many implications. A belief in God and an Afterlife resets the person's priorities, and redefines his measure of success and failure. The person would not limit his goals to his material achievements, and his concerns would not be limited to his own individual benefit anymore. Hence, one would understand the statement made by the Prophet's ﷺ companion Haraam Bin Malhaan when he was stabbed by a spear:

فُزْتُ وَرَبِّ الْكَعْبَةِ

I won, by the Lord of the Kaabah.[10]

Despite his fatal wound, he considered himself the victor, for his life was taken away for the sake of God. He understood success to encompass the Afterlife as well, as defined in the Qur'an:

$$\text{فَمَن زُحْزِحَ عَنِ النَّارِ وَأُدْخِلَ الْجَنَّةَ فَقَدْ فَازَ وَمَا الْحَيَاةُ الدُّنْيَا إِلاَّ مَتَاعُ الْغُرُورِ}$$

Only he who is saved far from the Fire and admitted to the Garden will have attained the object (of Life): For the life of this world is but goods and chattels of deception[11].

While belief in God used to come naturally to the earlier generations, it does not anymore. Today's world is largely dominated by materialism and swamped with a plethora of ideologies. Particularly in the West, everyone is encouraged to question and challenge everything. It is exalted and considered progressive where one would refuse to conform to the prevailing ways. Hence, it becomes increasingly difficult to *simply submit.*

Furthermore, the prevalent ideologies are not just present, they are continuously pressed onto everyone, and very effectively. With today's easy and universal access to the media, everyone has access to everyone else—and everyone proselytizes. Every ideology, correct or erroneous, has become highly visible, catching every receptive ear and mind. As a result of this constant bombardment, many people's minds and thinking get repeatedly hijacked by people of politics, people of science, as well as people of religion. Inevitably, this adds to people's confusion and doubts, further eroding their belief.

The above concerns are particularly serious for the younger generations. Young people do not naturally possess the skills to tolerate contradictions, which some adults have so well mastered. I have known personally some older scientists who consider themselves devout Christians, yet at the same time they believe in some scientific theories that contradict very clearly their own belief. They do not have any trouble tolerating the contradictions.

With a fair amount of experience in youth counseling (more than 20 years), I have yet to see a single youth with that quality.

To the self-honest young mind, everything has to be clearly defined. This quality brings the youth clarity and honesty; they are incapable of tolerating confusion and hypocrisy. But this same quality can be a bane when it comes to basic belief. A youth might easily reject a whole belief or ideology if it appears (to him/her) to contain confusing or incomprehensible elements.

The purpose of this book is to provide some insights into the role of the mind in establishing belief, to shed some light on the contradicting thoughts facing belief, and to provide some guidance to seekers of the truth. Though I will mention some general concepts that may apply to other religions, this study will be referring largely to the Islamic concept of belief. Still, people of other religions might find some of the discussions useful.

I did examine much of the existent literature on the topic. It should be noted that while the existent Islamic work is immensely valuable, it does not always address some of the present concerns, particularly those of the young Muslims living in the West. Moreover, most of the Islamic literature is written in Arabic and, as such, is not accessible to the non-Arabic speaker.

The book will address the following topics:
1) The nature of belief
2) Means to establish belief
3) The role of the mind in developing faith
4) Insights about doubt and atheism
5) The relationship and interface between science and religion

يَا أَيُّهَا الَّذِينَ آمَنُوا

O Ye who believe!

THE REALITY OF EEMAN

1. Definitions

It has been the habit of Islamic writers to define terms before discussing certain topics, to ensure that the appellations have the same indications for the writer and the reader. This ensures that everything is clarified and ideas are conveyed to the reader accurately. Hence, some basic terms will be defined below.

It is also essential to note that Islamic knowledge is closely linked to the Arabic language. Terms and concepts are defined in light of the Qur'an and the sayings of the Prophet ﷺ (referred to as "Hadeeth" moving forward), both of which were written in Arabic. A simple translation of a technical Islamic term from Arabic into just ONE English term without explanation would not be sufficient. It can lead to all sorts of erroneous conclusions and can easily convey the wrong meaning. Oftentimes, translations add new meanings and/or omit some other meanings from the original text.

One relevant example would be the translation of the word "وَلِيّ" (Waliy) in the Qur'an into "friend" by Abdullah Yusuf Ali[1] in his English translation of the Qur'an. The implications of such erroneous translations are obvious in translating such verses as 28 in Surat Aal-Imran:

---

[1] Abdullah Yusuf Ali (1872 – 1953) was an Indian Islamic scholar who translated the Qur'an into English. His translation of the Qur'an is one of the most widely known and used in the English-speaking world.

لاَ يَتَّخِذِ الْمُؤْمِنُونَ الْكَافِرِينَ أَوْلِيَاءَ مِن دُونِ الْمُؤْمِنِينَ وَمَن يَفْعَلْ ذَلِكَ فَلَيْسَ مِنَ اللَّهِ فِي شَيْءٍ إِلاَّ أَن تَتَّقُواْ مِنْهُمْ تُقَاةً وَيُحَذِّرُكُمُ اللَّهُ نَفْسَهُ وَإِلَى اللَّهِ الْمَصِيرُ

Translated by Yusuf Ali as:

*Let not the believers take for **friends** or helpers Unbelievers rather than believers: if any do that, in nothing will there be help from Allah; except by way of precaution, that ye may Guard yourselves from them. But Allah cautions you (To remember) Himself; for the final goal is to Allah.*

The Arabic verse calls upon the believers not to take the unbelievers as "Awliyaa" (plural of "Waliy"), translated by Yusuf Ali as "Friends." However, the word "Waliy" generally refers to someone who was given loyalty or guardianship over something/someone, and a "close friend" is only one of its many specific meanings.[12] Translating "Waliy" into simply "friend" would have some significant implications, particularly for the Muslim living in the West, among non-Muslims. According to Ali's translation, a Muslim should not take non-Muslims as friends. Yet the Prophet ﷺ did have friends among "unbelievers" (e.g., Hakim Bin Huzam [before he embraced Islam], Al-Mut`am Bin `Ady). Hence, a simple translation is sometimes insufficient.

A more complicated example is the translation of "القضاء والقدر" (Al-Qadaa' Wal-Qadar) into simply "destiny" or "pre-destiny." These mistranslations would convey a completely erroneous meaning: that the person has no ability to decide his course in life, and that everything is "pre-set" for him. The Qur'an indicates the opposite of that, stating that the person makes his own choice:

إِنَّا هَدَيْنَاهُ السَّبِيلَ إِمَّا شَاكِرًا وَإِمَّا كَفُورًا

We showed him the Way: whether he be grateful or ungrateful (rests on his will)[13].

24

"Al-Qadaa' Wal-Qadar" is a more complex concept that describes the knowledge, wisdom, and power of God. It is explained at length by the scholars of Aqeedah.[2,14]

Thus, technical Islamic terms and idioms have to be defined in light of their linguistic Arabic origins and their usage in the Qur'an and Hadeeth. I will explain the Arabic term, then choose a particular English translation to refer to it. We may not use the English word with all of its denotations and connotations, but rather with the specific meaning that we will define, in light of the Islamic idiomatic meaning.

A. Al-Fitrah

Al-Fitrah is defined in "Mu`jam Al-Ma`ani Al-Jame`" (معجم المعاني الجامع) as the initial state of creation of a person.[15] It is the collection of attributes and tendencies the human was initially given. The Qur'an states:

$$\text{إلاَّ الَّذِي فَطَرَنِي فَإِنَّهُ سَيَهْدِينِ}$$

Except for He who created me; and indeed, He will guide me[16].

$$\text{فَأَقِمْ وَجْهَكَ لِلدِّينِ حَنِيفًا فِطْرَةَ اللَّهِ الَّتِي فَطَرَ النَّاسَ عَلَيْهَا لا تَبْدِيلَ لِخَلْقِ}$$
$$\text{اللَّهِ ذَلِكَ الدِّينُ الْقَيِّمُ وَلَكِنَّ أَكْثَرَ النَّاسِ لا يَعْلَمُونَ}$$

So direct your face toward the religion, inclining to truth. [Adhere to] the fitrah of Allah upon which He has created [all] people. No change should there be in the creation of Allah. That is the correct religion, but most of the people do not know.[17]

Ibn Abbas said, "I didn't understand the meaning 'Faatir of Heavens' (فاطر السموات), as mentioned in the Qur'an, until two

---

[2] Abdul-Rahman Hassan Habannakah Al-Midani (عبد الرحمن حسن حبنكة الميداني; 1927 – 2004) was a Syrian Islamic scholar and writer who counts many scholars among his students, such as Muhammad Sa`eed Ramadan Al-Buti.

Bedouins came to me to resolve a dispute about a well. One said, 'I am the Faatir of the well,' meaning 'I started the well.'"[3]

The same meaning is repeated in the hadeeth where the Prophet ﷺ said:

$$كُلُّ مَوْلُودٍ يُولَدُ عَلَى الْفِطْرَةِ فَأَبَوَاهُ يُهَوِّدَانِهِ أَوْ يُنَصِّرَانِهِ أَوْ يُمَجِّسَانِهِ$$

Every child is born in the state of the Fitrah; his parents make him a Majusee, a Christian, or a Jew.[18]

Ibn Taimyah[4,19] wrote: "When we say a person is born on the state of the Fitrah or created *Haneef*, it does not mean that he came out of the womb of his mother knowing about this religion, for Allah brought us out of the womb of our mothers not knowing anything. Rather, his Fitrah gives him a natural inclination toward the religion of Al-Islam, to know it and to love it. And it is wrong to think that humans are created void of any knowledge or tendency to reject; rather, the Fitrah naturally pushes one in one direction, that of Eeman, and not in the direction of Kufr."[20]

Thus, Al-Fitrah can be defined as *the way a person is created, with his original and pure state of being.*

B. Eeman

A major word, most relevant to this study is "Eeman" (إيمان). The Qur'anic text indicates that:

1) Eeman is not just a verbal declaration:

---

[3] ابن كثير، أبو الفداء عماد الدين إسماعيل. 2002. « تفسير القرآن العظيم.» الجزء السابع. دار طيبة، الرياض، المملكة العربية السعودية
(Title in English: The Interpretation of the Great Qur'an).

[4] Taqiyyuddin Ahmad Ibn Taymiyyah (تقي الدين أحمد ابن تيمية), 1323 – 1263, known as simply Ibn Taymiyyah, was an Islamic scholar, commonly referred to as "Sheikhul-Islam." He was a member of the school founded by Ahmad bin Hanbal. Among his students was the famous Ibn Qayyim Al-Jawziyyah.

وَمِنَ النَّاسِ مَن يَقُولُ آمَنَّا بِاللَّهِ وَبِالْيَوْمِ الآخِرِ وَمَا هُم بِمُؤْمِنِينَ

*And of the people are some who **say**, "We believe in Allah and the Last Day," but they are not believers.*[21]

2) It is not just a physical action:

إِنَّ الْمُنَافِقِينَ يُخَادِعُونَ اللَّهَ وَهُوَ خَادِعُهُمْ وَإِذَا قَامُواْ إِلَى الصَّلاةِ قَامُواْ كُسَالَى يُرَاؤُونَ النَّاسَ وَلاَ يَذْكُرُونَ اللَّهَ إِلاَّ قَلِيلاً

Indeed, the hypocrites [think to] deceive Allah, but He is deceiving them. And when they stand for prayer, they stand lazily, showing (themselves to) the people and not remembering Allah except a little[22].

3) It is not just an intellectual conviction:

وَجَحَدُوا بِهَا وَاسْتَيْقَنَتْهَا أَنفُسُهُمْ ظُلْمًا وَعُلُوًّا فَانظُرْ كَيْفَ كَانَ عَاقِبَةُ الْمُفْسِدِينَ

And they rejected them, while their (inner) selves were convinced thereof, out of injustice and haughtiness. So see how was the end of the corrupters.[23]

Beyond any of the above individually, Eeman includes all, reaching the depths of the self, affecting all of its aspects. Intellectual conviction is a major component of Eeman:

قَالَتِ الأَعْرَابُ آمَنَّا قُل لَّمْ تُؤْمِنُوا وَلَكِن قُولُوا أَسْلَمْنَا وَلَمَّا يَدْخُلِ الإِيمَانُ فِي قُلُوبِكُمْ وَإِن تُطِيعُوا اللَّهَ وَرَسُولَهُ لا يَلِتْكُم مِّنْ أَعْمَالِكُمْ شَيْئًا إِنَّ اللَّهَ غَفُورٌ رَّحِيمٌ

The believers are only the ones who have believed in Allah and His Messenger and then doubt not but strive with their properties and their lives in the cause of Allah. It is those who are the truthful[24].

But the intellectual submission is followed by the submission of the heart and some concrete actions:

إِنَّمَا الْمُؤْمِنُونَ الَّذِينَ إِذَا ذُكِرَ اللَّهُ وَجِلَتْ قُلُوبُهُمْ وَإِذَا تُلِيَتْ عَلَيْهِمْ آيَاتُهُ زَادَتْهُمْ إِيمَانًا وَعَلَى رَبِّهِمْ يَتَوَكَّلُونَ

الَّذِينَ يُقِيمُونَ الصَّلاَةَ وَمِمَّا رَزَقْنَاهُمْ يُنفِقُونَ

أُوْلَئِكَ هُمُ الْمُؤْمِنُونَ حَقًّا لَّهُمْ دَرَجَاتٌ عِندَ رَبِّهِمْ وَمَغْفِرَةٌ وَرِزْقٌ كَرِيمٌ

The believers are only those who, when Allah is mentioned, their hearts become fearful, and when His verses are recited to them, it increases them in faith; and upon their Lord they rely. The ones who establish prayer, and from what We have provided them, they spend. Those are the believers, truly. For them are degrees [of high position] with their Lord and forgiveness and noble provision[25].

Like many of the Islamic terms, the word "Eeman" has a linguistic meaning, as well as an idiomatic meaning, called "Istilahi" (إصطلاحي).

Linguistically. Eeman is derived from the verb Aamana (آمن), which was originally A'mana (أءمن). It refers to conferring safety and tranquility, i.e., Amaan (أمان), upon someone and removing fear from them, as in the verse:

الَّذِي أَطْعَمَهُم مِّن جُوعٍ وَآمَنَهُم مِّنْ خَوْفٍ

Who (God) has fed them, (saving them) from hunger and made them safe, (saving them) from fear[26].

Based on that meaning, Eeman would refer to attaining safety and tranquility, which happens when the heart submits and accepts something as the truth.

"Eeman" may also refer to believing in something as being true (versus a lie), as in Surah Youssuf:

قَالُواْ يَا أَبَانَا إِنَّا ذَهَبْنَا نَسْتَبِقُ وَتَرَكْنَا يُوسُفَ عِندَ مَتَاعِنَا فَأَكَلَهُ الذِّئْبُ وَمَا أَنتَ بِمُؤْمِنٍ لَّنَا وَلَوْ كُنَّا صَادِقِينَ

They said, "O our father, indeed we went racing each other and left Joseph with our possessions, and a wolf ate him. But you would not believe us, even if we were truthful."[27]

Ibn Taimyah defined Eeman linguistically as not just "taking as true," or Tasdeeq (تصديق), but also as "accepting it," or Iqrar (إقرار). Thus, Eeman is not just "believing something," but further, it is "believing *in* something." And while the opposite of "Tasdeeq" is "Taktheeb" (taking something as a lie), the opposite of "Eeman" is "Kufr" (rejection).

This meaning is reflected in the hadeeth, where it mentions what the Prophet ﷺ used to say while circling around the Kaabah:

بسم الله ، والله أكبر ، اللهم إيماناً بك ، وتصديقاً بكتابك ، ووفاء بعهدك ، واتباعاً لسنة نبيك محمد. صلى الله عليه وسلم

*In the name of Allah, and Allah is greater; O Allah, doing it out of **believing in** (Eeman) You, and **believing** (Tasdeeq) Your book, and fulfilling the commitment to You, and following the guidance of Your prophet Muhammad, God's peace and blessing upon him.*

The hadeeth mentions both words "Tasdeeq" (believing something) and "Eeman" (believing *in* something), indicating they refer to different meanings. The hadeeth does have weakness in its authenticity, but still sheds light on the meaning of the words[28].

Thus, linguistically "Eeman" *is the heart's taking something as true AND accepting it and believing in it.*

Idiomatically/Istilah. Both the linguistic and the idiomatic dimensions of a word need to be taken into account when attempting to understand any Islamic concept. If the language tells us what "Eeman" generally means (i.e., believing in something/ accepting it), then the Islamic references tell us what specific meaning we should accept and what that entails.

These references include the understanding of the early Muslims because the Qur'an was revealed to them in the language they were using on a daily basis. We must first look at their understanding of a word/concept, then combine that with the linguistic indication. As such, their usage/understanding of the language becomes a controlling factor in our understanding of those words, even with the presence of a more general denotation of the words. Thus, when the Qur'an mentions "Eeman", it just mentions "The Eeman" (Al-Eeman), defined, with everyone already understanding what it refers to, due to the explanation already given by the Prophet ﷺ. So we don't just depend on the linguistic definition, or even use it as our main reference, if the idiomatic meaning is present and defined.

Eeman is defined as "a belief with certainty, and the complete acceptance of the Existence of God and His Attributes, and the belief in and acceptance of the Prophet ﷺ and all what he brought forth. That belief is followed by the verbal expression, and physical actions."[29] Thus, all three components are included: belief in the heart/mind, verbal expression, and physical action. Al-Shafi'i[5] defined Eeman as expression, action, and intention[30]. Some scholars define it as expression and action, with the "action" including "physical actions" as well as "actions of the heart" (in other words, belief in the heart).

Eeman is believed to increase through acts of obedience and diminish through acts of disobedience. Ibn Uthaimeen[6] stated that Eeman for Ahlus-Sunnah Wal-Jama'ah includes three components: belief in the heart, expression with the tongue, and action with the senses. As such, Eeman would increase and decreases in 2 ways: *1)* through change in the "belief" in the heart, when the level of conviction increases or decreases, i.e., with the strength of belief, and *2)* through change in the actions of the

---

[5]Abu Abdillah Muhammad Bin Idris Al-Shafi'i (Arabic: ابو عبدالله محمد بن إدريس الشافعيّ: 767 – 820) A Muslim jurist, one of the four great Imams of which a legacy on juridical matters and teaching eventually led to the Shafi'i school of fiqh (or Madh'hab) named after him. Hence he is often called Imam al-Shafi'i.

[6] Abu 'Abd Allah Bin Saalih Bin Al-Uthaymeen (Arabic: عبد الله بن صالح بن العثيمين: 1925 – 2001) was one of the most prominent Islamic scholars of his time. Along with Abd Al-Aziz Bin Baz, he was considered one of the two leading representatives of the conservative Saudi Arabian religious establishment.

senses, when the person performs various acts of obedience or disobedience.[31]

C. Al-Yaqeen

Al-Yaqeen (اليقين) is defined as a belief in something with certainty, a certitude. The word is mentioned in the Qur'an as in the following verses:

$$وَجَحَدُوا بِهَا وَاسْتَيْقَنَتْهَا أَنفُسُهُمْ ظُلْمًا وَعُلُوًّا$$

And they rejected those Signs in iniquity and arrogance, though their souls were convinced thereof[32]

$$وَاعْبُدْ رَبَّكَ حَتَّى يَأْتِيَكَ الْيَقِينُ$$

And worship thy Lord until there come unto thee the Hour of Certainty.[33]

Scholars often describe Eeman to include as a belief with "certainty that excludes doubt" (اليقين المنافي للشك), based on several texts, such as in the authentic hadeeth, where the Prophet ﷺ said:

$$أشهد أن لا إله إلا الله وأني رسول الله ، لا يلقى الله بهما عبد غير شاك فيهما إلا دخل الجنة$$

"I bear witness that there is not god but Allah and that I am the messenger of Allah", anyone meeting Allah with it without having doubt in it, shall enter Al-Jannah.[34]

This hadeeth, however, seems to describe the ideal. I do maintain that belief does not go up from zero to 100% (i.e., complete certainty) and stay there. It mostly fluctuates between the two and does not remain at the same strength all the time. But even when such belief does not reach certainty, it cannot be considered nullified. Some of our beliefs might not reach certainty due to their nature, and might remain in the "most likely" category; some certitude may not be reached so long as the person is alive.

This might be one of the meanings of verse 99 in Surat Al-Hijr, quoted above, "*And worship thy Lord until there come unto thee the Hour of Certainty.*" "Certainty" in that verse is commonly believed to refer to "death," and the verse is commonly translated as "worship your Lord until you die." But there could be another possible interpretation where "certainty" does not refer to "death," but rather that "death" **brings** "certainty." Then another way of translating the verse would be, "Worship your Lord until your belief reaches certainty, which is brought forth by death." Certainty and doubt will be discussed at length in later chapters.

That which is categorized as "certitude" (يقين) itself may be of different types and levels:

Certitude of Knowledge (CK), i.e., Ilmul-Yaqeen (علم اليقين): reached through relayed information, analysis and deduction.

Certitude of Sight (CS): Ainul-Yaqeen (عين اليقين): reached through direct perception.

Certitude of Truth (CT): Haqqul-Yaqeen (حق اليقين): reached through direct perception, confirmed and affirmed by direct experience.

Ibn Qayyim Al-Jawziyyah[7,35] stated that CK is reached through transmitted information. Then once the validity of this information is manifested to the eye or senses, it becomes CS. And when the person lives it and feels it, it becomes CT. For instance, our knowledge in this world of Al-Jannah (Paradise) can only reach CK. Once someone sees it on the Day of Judgment,

---

[7] Ibn Qayyim al-Jawziyyah, or simply Ibn Al-Qayyim, Muhammad Bin Abu Bakr (1292-1350؛ ابن قيّم الجوزية) was an Islamic jurist, commentator on the Qur'an and theologian. He is sometimes referred to by other Muslim scholars as "the scholar of the heart", given his extensive works pertaining to human behavior and ethics. He was the student of Sheikh Al-Islam Ibn Taymiyah.

it becomes CS. Then when that person enters it, her knowledge of Al-Jannah becomes CT.

Another example would be our knowledge of the ocean. When one reads about the ocean, the knowledge he acquires would lead to KC. Once he actually sees the ocean, that knowledge would move up to CS. Once that person starts swimming in the ocean, his knowledge reaches CT.

Believing in Allah's Existence can only reach CK. It can never be CS in this world. Yet, I believe that it may reach CT, without going through CS. It would happen by experiencing Him in one's heart, in daily life.

D. Eeman vs. Islam

Eeman and Islam are two expressions, which when mentioned separately, refer to the same meaning; and when mentioned together, refer to different meanings (إذا اجتمعا افترقا، وإذا افترقا اجتمعا).

When mentioned separately, Eeman and Islam refer to the whole comprehensive religion, as in the following verse and hadeeth:

إِنَّ الدِّينَ عِندَ اللَّهِ الإِسْلامُ

The Religion before Allah is Islam[36]

الإِيمَانُ بِضْعٌ وَسَبْعُونَ شُعْبَةً، أَوْ بِضْعٌ وَسِتُّونَ شُعْبَةً، أَفْضَلُهَا لَا إِلَهَ إِلَّا اللَّهُ، وَأَدْنَاهَا إِمَاطَةُ الأَذَى عَنِ الطَّرِيقِ، وَالْحَيَاءُ شُعْبَةٌ مِنَ الإِيمَانِ

Eeman is seventy and some sections, or sixty and some sections, the highest of which is the declaration if the oneness of Allah, the least of which is the removal of harm from people's way; and modesty is a segment of Eeman.[37]

When mentioned together, Eeman and Islam have different meanings, as indicated in the famous hadeeth narrated by Omar bin Al-Khattab in *Sahih Muslim*. In the hadeeth, Jibril came to

33

the Prophet ﷺ in the shape of a man and asked *"what is Islam?"*, to which the Prophet ﷺ replied *"to testify that there is no God but Allah and that Muhammad is His messenger, to establish prayer, to pay Zakaat, to fast Ramadan, and to perform pilgrimage if you have the means."* Then when asked *what Eeman is*, the Prophet ﷺ replied *"to believe in Allah, His Angels, His Books, His Messengers, the Last Day, and the Qadar, its good and its bad."* A similar distinction is made in Surat Al-Hujuraat:

$$\text{قَالَتِ الأَعْرَابُ آمَنَّا قُل لَّمْ تُؤْمِنُوا وَلَكِن قُولُوا أَسْلَمْنَا وَلَمَّا يَدْخُلِ الإِيمَانُ فِي قُلُوبِكُمْ}$$

The desert Arabs say, "We have Eeman." Say, "Ye have no Eeman; but you say, 'We have Islam (submitted our wills to Allah),' For not yet has Eeman entered your hearts.[38]

Islam would then refer to the physical action and submission to the will of God, while Eeman would refer to the internal deeds, the heart's belief in the entire basic creed of the religion. "Eeman" and "belief" will be used interchangeably in this book.

E. Ghaib and Shahadah

In the context of Eeman, the world can be divided into two components: "the World of Shahadah" (عالَم الشهادة), and "the World of Ghaib" (عالَم الغَيب). The World of Shahadah is that which is witnessed, experienced first-hand, and/or perceived through the senses. The World of Ghaib is that which is not experienced directly, due to spatial or temporal limitations, or due to its metaphysical nature.

A temporal or spatial Ghaib is only temporary or conditional. Once the spatial or temporal barriers are removed, it becomes Shahadah. For example, events occurring in another country are part of the spatial Ghaib; but, once one travels to that country, the spatial limitations are removed and those events become part of the World of Shahadah. Similarly, what is in the womb of a pregnant woman is a Ghaib; once an ultrasound scan is performed or the woman gives birth, it becomes a Shahadah.

The metaphysical Ghaib will remain as such so long as the person is in this world. Knowledge of this kind of Ghaib is exclusively with God. The distinction is made in Surat Loqman:

$$\text{إِنَّ اللَّهَ عِندَهُ عِلْمُ السَّاعَةِ وَيُنَزِّلُ الْغَيْثَ وَيَعْلَمُ مَا فِي الْأَرْحَامِ}$$

Verily the knowledge of the Hour is with Allah (alone). It is He Who sends down rain, and He knows what is in the wombs.[39]

The verse mentions the "knowledge of the Hour" and "what is in the wombs." With the former, He used the expression ***"the knowledge … is with Allah,"*** since it is exclusively with God, without anyone else having access to such knowledge. But since the latter knowledge can be acquired by humans as well (through various tools), He used the expression ***"He knows,"*** which indicates that others are not excluded from knowing it as well.

Another example of something that will remain a Ghaib is knowledge related to the spirit (الرُّوح). Allah says:

$$\text{وَيَسْأَلُونَكَ عَنِ الرُّوحِ قُلِ الرُّوحُ مِنْ أَمْرِ رَبِّي وَمَا أُوتِيتُم مِّنَ الْعِلْمِ إِلاَّ قَلِيلاً}$$

They ask thee concerning the Spirit. Say: "The Spirit (cometh) by command of my Lord: of knowledge it is only a little that is communicated to you, (O men!)"[40]

Knowledge of the Afterlife, of the Angels, and of the Jins are other examples of the metaphysical Ghaib. As will be explained later, our belief in the metaphysical Ghaib might still be very strong, despite its being a complete Ghaib. However, the path to developing such belief is different than that used for belief in Shahadah. Belief in something of a metaphysical nature will always be qualitatively different than a belief in something experienced directly, of a Shahadah nature. One cannot expect to directly perceive something metaphysical in nature.

2. Paths to Establish a Belief

Since belief is of pivotal importance in one's life, ensuring the correctness of the belief becomes essential. A sound path to establish a belief leads itself to a sound belief, while a false path may lead to an erroneous belief. There are six paths to establish any form of belief:

A. Through Direct Perception

Perceiving something through the five senses (seeing, hearing, smelling, touching, tasting), directly or through the use of some tools, leads to a direct belief in its existence. For instance, seeing an object indicates its presence and leads to a belief in its existence. If we perceive it, it exists, and our belief in its existence would be certain, since our senses are normally accurate. Such belief can reach CS.

Our senses are accurate in what they perceive directly, but our senses do have limitations in their perceptive abilities. We know that our eyes do not show us everything. For instance, there are minute microorganisms floating around us that we do not perceive. Similarly, we know that there are certain sound frequencies travelling all around us that we cannot hear. Thus, if we see something, it does exist, but the opposite is not true; if we do not see it, it does not mean it does not exist. The same can be said about the other senses.

Our senses are our windows to the world around us, our perception, and our impression of the world are shaped by our perception. Thus, someone with limited senses will have a limited perception of the world. Someone who lacks one of the senses altogether (sight, hearing, smelling) will have an incomplete perception of the world as we know it. For instance, a person without a sense of smell would never have guessed, on his own, the existence of that "dimension" of the world.

For that person, just like for every person, the world is not limited to what he perceives through his senses. There are certainly aspects of our world that we do not know exist, simply because we do not have the senses to perceive them. And just as it would be absurd for us if the above-mentioned person denies the existence of "smell," it is equally absurd for us to deny the existence of something simply because we cannot perceive it through our senses. We are continuously learning of new aspects of our world that we did not perceive previously, simply because we did not have the tools or the senses to detect them. For instance, insects perceive the world differently than we do, often sensing certain stimuli that humans did not even know existed. Therefore, we cannot rule out the existence of something, even the existence of other dimensions, simply because we cannot perceive them.

Thus, our senses could help us develop a belief in aspects of the "World of Shahadah," but not of the "World of Ghaib." That belief would have to come through different paths. Our senses only give us a *partial* perception of the whole reality.

B. Indirectly, Through Analysis/Deduction

Certain beliefs cannot be reached through direct perception, but rather indirectly through a cognitive/intellectual process. We may develop a belief in something through analysis, without having to perceive it directly. For example, by seeing a well-manicured front yard, we may conclude the existence of a gardener, even if we have not seen or heard that person.

By being indirect compared to "direct perception," this method introduces a possibility of error. Hence with this method, attention needs to be placed on the *process* of analysis that leads to the conclusion, and eventually to a belief. A flawed thinking process leads to a false belief.

As we have concluded earlier about our senses, it is important to note that the mind itself has limitations. A major limitation is due to the fact that our own world is limited by time and space. Our life is confined to these two dimensions, and we always think of "when" and "where." That is why people are still struggling to comprehend the notion of the finite vs. infinite nature of this universe. We cannot imagine an infinite universe, and we cannot imagine it could have an end. Our inability to comprehend something, however, does not indicate its inexistence. Our mind may just not comprehend it. Al- Imam Al-Shafi`i said, "the human mind has limits at which it stops, just as sight has limits." Just as sight is limited by distance and resolution, so does the mind have limits, beyond which it cannot provide answers on its own.

A major component to this path to establish belief is imagination. It allows us to see (with our mind) certain things that our senses do not show us, and which we arrive at through our analysis. But imagination itself, rich as it is, also has limitations. One can only imagine something which he has experienced before. For example, any shape we imagine is based, even if just partially, on something we have seen earlier. This constraint obviously adds to the limitation of the mind and of this method, as a means to establish a belief.

Thus, one cannot, through his intellectual ability alone, pass a judgment concerning something he has never been exposed to, or told about. He cannot discover new realities purely through analysis, without being given specific information about it. Therefore, a belief in something metaphysical in nature (*Al-Jannah*, Angels, *Jins*, etc.) needs to follow a different path than merely an intellectual analysis. The same can be said about determining the higher purpose of one's own existence. This would have to be given to us, as described in the next path.

But despite the limitations of this method, it remains an essential tool to judge the validity of some beliefs. Though some beliefs can only be established through relayed information, an intellectual examination can establish its correctness or fallacy. It can determine if something is impossible (even if metaphysical), if it contradicts some basic logic such as containing two mutually exclusive aspects.

Our belief through this method can reach CK.

Something to be noted about the first two paths to establish a belief: a belief is not established simply by perceiving something or coming to an intellectual conclusion about it. This would only lead to information or an opinion about it. Perception and intellectual conviction have to be *repeated* consistently for something to become a "belief." Our knowledge about something is first perceived, then when repeated, moves from perception into knowledge. Then through time, repetition, and reinforcement, it changes from perception and knowledge into belief. For example, one might see a nice piece of art and conclude that the artist is talented. This is only an opinion. But when she sees other works by the same artist, using different media and techniques, she will conclude solidly that the artist is truly gifted. This is firm knowledge. Then if she gets exposed to the artist further, studying his techniques, joining his discussion platforms and fan page, etc., knowledge concerning this artist becomes a belief.

C. Through Relayed Information

We may develop a belief in something, even strongly, simply because it was conveyed and described to us by a trustworthy source of information. Much of our belief is actually developed through these means: historical events that we have not witnessed (e.g., World Wars), geographical places we have not visited, events occurring in different places, scientific findings, etc. We believe in their existence because information concerning them reached us through independent credible sources, and our belief in their existence may reach the level of certainty though we have not witnessed them directly.

Efforts in this method would be focused on establishing the credibility of the messenger. This was done by the Prophet's ﷺ close companion Abu Bakr when he was informed by some messenger of the Prophet's ﷺ claim concerning his Israa' and Mi`raaj. Abu Bakr trusted the Prophet ﷺ, but the credibility of that messenger/ source is questionable. Hence he made the statement:

والله لئن كان قاله لقد صدق

*By Allah, **if** he (the Prophet) had said it, then indeed he said the truth.*[41]

Certain beliefs cannot be reached through the first two methods due to certain limitations that are spatial, temporal, or metaphysical. They cannot be perceived directly, and the mind cannot reach them on its own. Here, relayed information would be the only means to learn about them. Most of the metaphysical Ghaib falls under this category. Here, the duty of God's Prophets and Messengers would be to convey such knowledge to us, which reached them themselves through the divine revelation.

Despite such knowledge being reached through an intermediary messenger, it could lead to a belief as strong as a belief established through direct perception. A belief developed through this method can reach CK.

D. Answering the Call of the Fitrah

As indicated earlier by Ibn Taymiyah, one's Fitrah gives him the natural tendency to believe. People have a natural longing for the Divine. But, though answering the call of the Fitrah may lead to a strong belief, it does not lead to objective certainty. It is not completely accurate, as one's Fitrah is not normally preserved in its initial pure state. It does get "contaminated" through excessive exposure to materialism and various ideologies.

Although the Fitrah gets covered by various layers of confusion, it resurfaces in times of extreme difficulties as described in the following verse:

هُوَ الَّذِي يُسَيِّرُكُمْ فِي الْبَرِّ وَالْبَحْرِ حَتَّى إِذَا كُنتُمْ فِي الْفُلْكِ وَجَرَيْنَ بِهِم بِرِيحٍ طَيِّبَةٍ وَفَرِحُواْ بِهَا جَاءَتْهَا رِيحٌ عَاصِفٌ وَجَاءَهُمُ الْمَوْجُ مِن كُلِّ مَكَانٍ وَظَنُّواْ أَنَّهُمْ أُحِيطَ بِهِمْ دَعَوُاْ اللَّهَ مُخْلِصِينَ لَهُ الدِّينَ لَئِنْ أَنجَيْتَنَا مِنْ هَذِهِ لَنَكُونَنَّ مِنَ الشَّاكِرِينَ

He it is Who enableth you to traverse through land and sea; so that ye even board ships;- they sail with them with a favourable wind, and they rejoice thereat; then comes a stormy wind and the waves come to them from all sides, and they think they are being overwhelmed: they cry unto Allah, sincerely offering (their) duty unto Him saying, "If thou dost deliver us from this, we shall truly show our gratitude!"[42]

E. Through Basic Feelings

Related to the previous method, a feeling could lead to a belief. Sometimes merely feeling something is an indicator of its existence, whether we can articulate the proof or not. For instance, a mother may feel the presence of her child, without being told about it.

A feeling becomes an even stronger indicator if it is shared *independently* by many people. This is true concerning our feeling of the Existence of a Creator, of God. People share that feeling, across various cultures, religions, geographical regions, and time periods. As such, that feeling becomes a strong indicator of its validity.

Clearly, a belief established merely based on feelings does not stand on its own, as feelings can be influenced by so many factors such as biases, personality traits, background, etc. It does not lead to certainty, but it still can give a good indicator.

F. Following the Course of Earlier Generations

Some people simply follow the course of previous generations or other people in choosing a belief, without any other consideration. Clearly, such a path can be erroneous, unless verified. Yet, uncertain as it is, it appears to be the way most people choose their belief. People tend to mostly adopt the belief they grew up with. It maintains familiarity and gives comfort. The Prophet ﷺ pointed it out in the hadeeth:

كُلُّ مَوْلُودٍ يُولَدُ عَلَى الْفِطْرَةِ فَأَبَوَاهُ يُهَوِّدَانِهِ أَوْ يُنَصِّرَانِهِ أَوْ يُمَجِّسَانِهِ

Every child is born on the state of the Fitrah; his parents make him a Majoosee, a Christian, or a Jew.[43]

Also, Allah states in the Qur'an that some people blindly adhere to the religion of their forefathers:

وَإِذَا قِيلَ لَهُمُ اتَّبِعُوا مَا أَنزَلَ اللَّهُ قَالُواْ بَلْ نَتَّبِعُ مَا أَلْفَيْنَا عَلَيْهِ آبَاءَنَا أَوَلَوْ كَانَ آبَاؤُهُمْ لاَ يَعْقِلُونَ شَيْئًا وَلاَ يَهْتَدُونَ

When it is said to them: "Follow what Allah hath revealed:" They say: "Nay! we shall follow the ways of our fathers." What! even though their fathers were void of wisdom and guidance?[44]

Obviously, these texts do NOT call one to reject something, simply because it was followed by the early generations. It may be that those generations were actually on the true path. The call is rather for the person to learn about the truth himself and develop that belief further, based on the methods described above, namely 1 through 3.

ذَٰلِكَ ٱلْكِتَٰبُ لَا رَيْبَ فِيهِ هُدًى لِّلْمُتَّقِينَ

This is the Book, in it is guidance sure,
without doubt, to those who are pious!

ADDRESSING DOUBT

Faith doesn't mean you never doubt. It only means you never act upon your doubts.
— Orson Scott Card, *Sarah*

1. Defining Doubt

Different dictionaries define doubt mostly similarly, with slight variations:

- *A feeling of **uncertainty** about the truth, reality, or nature of something.*[45]
- *The state of being **uncertain** about the truth or reliability of something. The condition of being **unsettled** or **unresolved**.*[46]
- *A feeling of not knowing what to believe or what to do, or the condition of being **uncertain**.*[47]

The Arabic word for "doubt" is "Shakk" (شَك). Similarly to the English definitions given above, "doubt" in the Arabic language, and as used in the Qur'an, does not always refer to a position where two options are equally likely, but more generally, to anything short of certainty.

- It can refer to a lack of knowledge about something, and to various levels of uncertainty.[48]

وَإِنَّ الَّذِينَ اخْتَلَفُواْ فِيهِ لَفِي شَكٍّ مِّنْهُ مَا لَهُم بِهِ مِنْ عِلْمٍ إِلاَّ اتِّبَاعَ الظَّنِّ

45

- *... and those who differ therein are full of doubts, with no (certain) knowledge, but only conjecture to follow[49]*

In this verse, "doubt" describes a state of lack of conclusive knowledge. The same meaning is repeated in this verse:

بَلِ ادَّارَكَ عِلْمُهُمْ فِي الآخِرَةِ بَلْ هُمْ فِي شَكٍّ مِّنْهَا بَلْ هُم مِّنْهَا عَمُونَ

Still less can their knowledge comprehend the Hereafter: Nay, they are in doubt and uncertainty concerning it; nay, they are blind thereunto![50]

The verse describes people who are in a state of doubt due to their lack of knowledge.

- Sometimes "doubt" refers directly to a lack of knowledge as in this verse:

فَإِن كُنتَ فِي شَكٍّ مِّمَّا أَنزَلْنَا إِلَيْكَ فَاسْأَلِ الَّذِينَ يَقْرَؤُونَ الْكِتَابَ مِن قَبْلِكَ

- If thou wert in doubt as to what We have revealed unto thee, then ask those who have been reading the Book from before thee.[51]

Meaning if you, O Muhammad, had no knowledge (i.e., were in a state of "doubt") of what the Torah and the Bible said about you, then ask the People of the Book, for they have that knowledge in their Books; so, they know you as well as they know their children.

- Another meaning of "doubt" is "rejection," as presented in this verse:

قُلْ يَا أَيُّهَا النَّاسُ إِن كُنتُمْ فِي شَكٍّ مِّن دِينِي فَلاَ أَعْبُدُ الَّذِينَ تَعْبُدُونَ مِن دُونِ اللَّهِ وَلَكِنْ أَعْبُدُ اللَّهَ الَّذِي يَتَوَفَّاكُمْ وَأُمِرْتُ أَنْ أَكُونَ مِنَ الْمُؤْمِنِينَ

- Say: "O ye men! If ye are in doubt as to my religion, (behold!) I worship not what ye worship, other than Allah. But I worship Allah - Who will take your souls (at death): I am commanded to be (in the ranks) of the Believers[52]

In this verse, "doubt" does not refer to hesitation or lack of certainty, but rather to rejection, for the people mentioned in this verse had rejected the Prophet ﷺ and his religion.

46

Therefore, we may define "doubt" as *an uncertain state of mind, ranging between minor hesitation and confusion.* It can refer to a minor level of uncertainty, as well as to complete suspension between two contradictory propositions, where a person is equally leaning toward both directions. Doubt can occur either when the evidence *for* and *against* is so equally balanced as to render decision impossible, or when sufficient evidence is totally or partially absent on either side.

Doubt can affect various aspects of one's life. It can impede one's decision-making by creating excessive hesitation among different courses of action. Doubt can even destroy relationships by poisoning one's ideas about other people. Some people pathologically doubt other people's motives and behaviors.

Doubt could also affect one's religious worships and practices, and, more seriously, belief itself. Different aspects of one's belief may be contaminated to various degrees with serious doubts. People may doubt God's Existence, justice, mercy, and so on. Central to all of our discussion is obviously the question of the belief in God and His Existence, and our analysis will focus on that aspect in particular.

God's Existence is supposed to be one of the basic and obvious realities that the Fitrah should guide one toward. The human has a natural tendency to believe in some divine super power that is controlling the universe and a natural tendency to seek it (Him).

قَالَتْ رُسُلُهُمْ أَفِي اللَّهِ شَكٌّ فَاطِرِ السَّمَاوَاتِ وَالْأَرْضِ

Their apostles said: "Is there a doubt about Allah, The Creator of the heavens and the earth?[33]

But with the spread of materialism and various ideologies, doubt sets in. We get the *illusion* that the evidence *for* and *against* God's Existence is so equally balanced as to render decision impossible. Addressing doubt and the validity of the evidence in light of the Qur'an and the mind becomes pivotal.

47

2. Doubt vs. Satanic Whispers

Doubt is often confounded with satanic whispers (وَسْوَسَة الشّيطان), and in most of Islamic literature the two are combined and discussed as one. They are mostly addressed as "satanic whispers," with real doubt commonly ignored. Making the distinction between the two is essential, as they should be dealt with differently.

Interestingly, the Qur'anic text itself makes a clear distinction between the two. While doubt is defined as lack of knowledge or caused by lack of knowledge, as presented earlier, satanic whispers refer to internal suggestions and thoughts, as in the following verses.

وَإِمَّا يَنزَغَنَّكَ مِنَ الشَّيْطَانِ نَزْغٌ فَاسْتَعِذْ بِاللّهِ إِنَّهُ سَمِيعٌ عَلِيمٌ

إِنَّ الَّذِينَ اتَّقَوا إِذَا مَسَّهُمْ طَائِفٌ مِّنَ الشَّيْطَانِ تَذَكَّرُواْ فَإِذَا هُم مُّبْصِرُون

If a suggestion from Satan assails thy (mind), seek refuge with Allah, for He heareth and knoweth (all things).
Those who fear Allah, when a thought of evil from Satan assaults them, bring Allah to remembrance, when lo! They see (aright).[54]

وَإِمَّا يَنزَغَنَّكَ مِنَ الشَّيْطَانِ نَزْغٌ فَاسْتَعِذْ بِاللَّهِ إِنَّهُ هُوَ السَّمِيعُ الْعَلِيمُ

And if (at any time) an incitement to discord is made to thee by the Evil One, seek refuge in Allah. He is the One Who hears and knows all things.[55]

These verses describe the satanic whispers as some passing thoughts that have no foundation, and thus are best ignored.

In a series of hadeeths, the Prophet ﷺ echoed the same meaning and described the satanic whispers in more detail. In a hadeeth narrated in *Sahih Bukhari*:

قال رسول الله – صلى الله عليه وسلم -: "يأتي الشيطان أحدكم فيقول: من خلق كذا؟ من خلق كذا؟ حتى يقول: من خلق ربك؟ فإذا بلغه فليستعذ بالله ولينتهِ"

The Prophet ﷺ said "The Devil comes to one of you and says 'Who created such, and who created such' until he says 'who created your Lord?' if someone find that in him, he should seek refuge in God and stop."[36]

In a version of the hadeeth narrated in *Sahih Muslim*, the Prophet said *"If someone finds any of that in him, he should say 'I believe in Allah.'"*[37]

("فمن وجد من ذلك شيئاً، فليقل: آمنت بالله").

There are other narrations of the hadeeth that repeat the same meaning, again indicating that such thoughts (i.e., the satanic whispers) should not be dignified.

In another hadeeth:

قال: جاء ناس من أصحاب النبي – صلى الله عليه وسلم – فسألوه: إنا نجد في أنفسنا ما يتعاظم أحدنا أن يتكلم به؟ قال: "وقد وجدتموه؟" قالوا: نعم قال : "ذاك صريح الإيمان"

Some of the Prophet's companions (S) came to him and asked him "sometimes we find in our hearts some thoughts that we consider too serious for us to mention." He said "you found that?" They said "yes." He said "this is the true faith."[38]

There are some other slightly different narrations, but they still convey the same meaning.

Based on the above texts, satanic whispers can be defined as *internal voices that make one hang onto some thoughts, that often have no basis, and that take over one's mind in the severe compulsive cases.* They can be considered as *baseless* and *recurring* doubts. Someone who "doubts" much may be actually afflicted by these whispers rather than real doubts, and his doubts may not be justified.

Satanic whispers can be associated with acts of worship where the person doubts every act related to cleanliness, worship, etc. They can also affect one's dealing with others where his social interactions are impacted by baseless fears and ill-feelings toward others.

And, as with real doubt, satanic whispers can also affect one's belief. Some people have a propensity to regularly doubt God's Existence, His wisdom, His justice, with such thoughts occurring uncontrollably and without any basis.

While a certain thought may be considered a legitimate doubt for one person, the *same exact thought* may be a mere baseless whisper for someone else. Consequently, while the legitimate doubt needs to be addressed and answered for the former, the satanic whisper of the same thought should be ignored for the latter.

Commenting on hadeeth Abi Hurairah quoted above ("Who created your Lord?"), Hassan Al-Banna accurately explained in his "Message of Beliefs"[59] ("رسالة العقائد") that this hadeeth illustrates how we project human attributes whenever we think about the Self of God (i.e., the problem of anthropomorphism). To restate his explanation differently, the question "Who created your Lord?" is based on the assumption that the Lord was actually created, and the one asking the question is trying to find out who created Him. So the problem is with the assumption on which the question is based, i.e., that God was created. Trying to find an answer to the question ignores the fact that the whole question itself is flawed with an invalid assumption.

But, interestingly, the Prophet ﷺ did not give an explanation as to why the question should not be asked. He just dismissed the whole question. Imam Al-Banna addressed the question as a real doubt; the Prophet ﷺ more correctly addressed it as a mere compulsive whisper. The Prophet ﷺ may have done so because the question is based on a flawed logic as explained above. Or, and I do believe it to be the case, because the Prophet ﷺ did not even want to dignify the whole thought, dismissing it as a mere satanic whisper. He just asked the companion to remind himself of his basic belief (say "I believe in Allah"), and that a passing thought (whisper) does not affect his established belief. I do believe this to be the best approach to deal with compulsive satanic whispers. Answering them would strengthen them and may just replace them with a stream of never ending questions. Ignoring them would eliminate them at the root.

Ignoring these thoughts becomes particularly important for some people who have a compulsive inclination for these thoughts and whispers. These recurring thoughts can cause them much agony, until eventually they start questioning their own faith. I have met several people who agonize over resisting these compulsive whispers, and wrongly label themselves as disbelievers or atheists. Yet these voices in their head were no more than passing thoughts that do NOT reflect their beliefs or convictions. Recognizing them for what they really are (mere whispers) keeps them trivial and insignificant. Dwelling on them makes them stronger. It is just like a child that compulsively imagines the shape of God after being told strictly that nothing resembles God.

The following is recommended in dealing with satanic whispers:

1. To recognize them as mere whispers.

$$\text{إِنَّ الَّذِينَ اتَّقَوا إِذَا مَسَّهُمْ طَائِفٌ مِّنَ الشَّيْطَانِ تَذَكَّرُوا فَإِذَا هُم مُّبْصِرُونَ}$$

2. *Those who fear Allah, when a thought of evil from Satan assaults them, bring Allah to remembrance, when lo! They see (aright)!*[60]

Recognizing the whispers for what they really are (compulsive thoughts and not real doubts) is important for knowing how to deal with them. The question then is how to tell if the thoughts in one's mind are real doubts or whispers.

Firstly, if these thoughts are baseless and recurring compulsively, then most likely they are whispers. Secondly, one needs to observe her overall tendencies, even in other areas. For example, if she tends to frequently and compulsively doubt her prayers and ablutions, then she is likely to be prone to satanic whispers. Her thoughts in other areas might be afflicted by the same, and her doubts might actually be mere whispers. The same might be true with people that hesitate much or have difficulties focusing in general. It would be interesting to examine the relationship between Attention Deficit Disorder and the propensity to doubts and satanic whispers.

3. To ignore them and not to dwell on them, as explained above.

4. To seek refuge in God from these thoughts, as God commands in the Qur'an.

$$\text{وَإِمَّا يَنزَغَنَّكَ مِنَ الشَّيْطَانِ نَزْغٌ فَاسْتَعِذْ بِاللّهِ إِنَّهُ سَمِيعٌ عَلِيمٌ}$$

5. If a suggestion from Satan assails thy (mind), seek refuge with Allah, for He heareth and knoweth (all things).[61]

$$\text{وَإِمَّا يَنزَغَنَّكَ مِنَ الشَّيْطَانِ نَزْغٌ فَاسْتَعِذْ بِاللّهِ إِنَّهُ هُوَ السَّمِيعُ الْعَلِيمُ}$$

6. And if (at any time) an incitement to discord is made to thee by the Evil One, seek refuge in Allah. He is the One Who hears and knows all things.[62]

7. To expect the reward for this struggle. Al-'Izz Bin Abdel-Salam[1] and others indicate that the person fighting satanic whispers would have the reward of the Mujahid (the one struggling for the sake of God), for he is fighting the enemy of God, i.e., Satan.

8. To keep in mind that the Muslim is NOT accountable for

---

1 Al-'Izz Bin Abdel-Salam 1262–1181 ;العز بن عبد السلام was an Islamic scholar and A Shafi'ee jurist, known as the "the Sultan of Scholars." He came to prominence during the Crusades toward the end of the Abbasid Caliphate. He was most known for his forceful pushing of the Muslim leaders to resist the invasion of the Moguls and Tatars.

these thoughts, so long as they stay as such, without following them with actions. In an agreed upon hadeeth, the Prophet ﷺ said:

إن الله عز وجل تجاوز لأمتي ما وسوست به وحدثت به أنفسها ما لم تعمل أو تتكلم به

9. Allah the Almighty has forgiven my Ummah for their satanic whispers and their self-talks, so long as they do not work or talk according to these thoughts.[63]

10. To avoid situations that are conducive to these whispers. For example, it would not be advisable for people that are prone to these whispers to engage in the likes of intellectual debates. Similarly, reading certain books that fuel these compulsive thoughts should be avoided.

3. Causes of Doubt

Doubt can be *positive* or *negative*.[64] In a positive doubt the evidence *for* and *against* is equally balanced, rendering decision impossible. On the other hand, a negative doubt arises from the absence of sufficient evidence on either side.

I believe, more significantly, doubt can be categorized as *direct* or *indirect*. A direct doubt is straight-related to the subject in doubt, due to confusing or missing evidence (positive or negative doubt). But doubt can be also indirect, not related to the proposition or idea itself but rather a reflection of another problem. The same can be said about extreme cases of doubt and atheism (please refer to the next section). For instance, someone may develop doubts about Islam because of a bad experience with the people of religion, or because of a vested interest in the "other option," and not because of the tenets of the religion.

Over the years, I met four young men who converted from Islam into Christianity. When visiting with them, they did all have some questions and issues with Islam which they presented to me. After thoroughly discussing these issues, they would all say that everything was now clarified, and their misconceptions of Islam were cleared. Yet still, they adhered to their conversion to the new faith. After further discussion with them and their families, I found out that the real factors leading to their conversion were not the questions they raised. In each of the four cases, there was a Christian woman the man was interested in, and she was the main factor that drew him into the new faith.

Direct doubt can be addressed by correcting misinformation where possible. An honest presentation of proof would clarify the issue and dissipate the confusion. The Qur'an continuously calls for a logical exchange and a kind discussion with others, while attempting to convey the truth and dissipate misconceptions.

ادْعُ إِلَى سَبِيلِ رَبِّكَ بِالْحِكْمَةِ وَالْمَوْعِظَةِ الْحَسَنَةِ وَجَادِلْهُم بِالَّتِي هِيَ أَحْسَنُ إِنَّ رَبَّكَ هُوَ أَعْلَمُ بِمَن ضَلَّ عَن سَبِيلِهِ وَهُوَ أَعْلَمُ بِالْمُهْتَدِينَ

Invite (all) to the Way of thy Lord with wisdom and beautiful preaching; and argue with them in ways that are best and most gracious: for thy Lord knoweth best, who have strayed from His Path, and who receive guidance.[65]

The Qur'an even goes further into calling one to go down to the level of his opponents, in order to help them see the truth:

قُلْ مَن يَرْزُقُكُم مِّنَ السَّمَاوَاتِ وَالْأَرْضِ قُلِ اللَّهُ وَإِنَّا أَوْ إِيَّاكُمْ لَعَلَى هُدًى أَوْ فِي ضَلَالٍ مُّبِينٍ

*Say: "Who gives you sustenance, from the heavens and the earth?" Say: "It is Allah. **And certain it is that either we or ye are on right guidance or in manifest error!"**[66]*

One example is the exchange between Prophet Abraham (S) and the king:

أَلَمْ تَرَ إِلَى الَّذِي حَاجَّ إِبْرَاهِيمَ فِي رَبِّهِ أَنْ آتَاهُ اللَّهُ الْمُلْكَ إِذْ قَالَ إِبْرَاهِيمُ رَبِّيَ الَّذِي يُحْيِي وَيُمِيتُ قَالَ أَنَا أُحْيِي وَأُمِيتُ قَالَ إِبْرَاهِيمُ فَإِنَّ اللَّهَ يَأْتِي بِالشَّمْسِ مِنَ الْمَشْرِقِ فَأْتِ بِهَا مِنَ الْمَغْرِبِ فَبُهِتَ الَّذِي كَفَرَ وَاللَّهُ لاَ يَهْدِي الْقَوْمَ الظَّالِمِينَ

Hast thou not turned thy vision to one who disputed with Abraham About his Lord, because Allah had granted him power? Abraham said: "My Lord is He Who giveth life and death." He said: "I give life and death". Said Abraham: "But it is Allah that causeth the sun to rise from the east: Do thou then cause him to rise from the West." Thus was he confounded who (in arrogance) rejected faith. Nor doth Allah Give guidance to a people unjust.[67]

In addressing the roots of doubt, Al-Jisr[68] stated something that confirms my personal observations, that the main cause of doubts is the inability to understand the Qadar (decree) of God in this world. Doubt arises from our failure to comprehend the events happening to us or to those around us. Why is there so much suffering in this world? Why do people earn differently, sometimes regardless of the efforts they exert? Why do some people seem to be favored far more than others? Why is my life so hard?... So many questions, all focusing on the same idea: *"Why do certain things happen in a certain way?"*

The problem with these questions is that we are trying to find answers that are unavailable, or, more correctly, inaccessible, simply because we do not have the complete story, not yet.

An examination of the story of Prophet Yussuf (Joseph) as mentioned in the Qur'an perfectly illustrates that point. Prophet Yussuf went through a series of hardships growing up. He was left in a well by his older brothers, picked up by some slave merchants, sold as a slave to Egypt's king, tempted by the king's wife, and imprisoned for years. Then, after these hardships, he was appointed the treasurer of Egypt, and the Surah further states that his reward in the Hereafter is even greater.

وَكَذَلِكَ مَكَّنَّا لِيُوسُفَ فِي الأَرْضِ يَتَبَوَّأُ مِنْهَا حَيْثُ يَشَاء نُصِيبُ بِرَحْمَتِنَا مَن نَّشَاء وَلاَ نُضِيعُ أَجْرَ الْمُحْسِنِينَ

وَلَأَجْرُ الآخِرَةِ خَيْرٌ لِّلَّذِينَ آمَنُواْ وَكَانُواْ يَتَّقُونَ

Thus did We give established power to Joseph in the land, to take possession therein as, when, or where he pleased. We bestow of our

Mercy on whom We please, and We suffer not, to be lost, the reward of those who do good.

But verily the reward of the Hereafter is the best, for those who believe, and are constant in righteousness.[69]

If we were to only see the young boy while being sold as a slave, we would not be able to understand "why" it is happening and would get the wrong impression. Here is a good, righteous boy, the son of a Prophet (Ya`coob [Jacob]), in a very dire situation. How could that happen? Obviously, having the complete story in our hands, we understand where this sad instance falls in the larger picture, and how the glorious end has to go through some difficult events. These hard events led to his eventual success in this world, and still bigger success and reward in the Hereafter. Now we understand "why."

Similarly, when one tries to understand the differences among people and the hardship in this world, he will be faced with a dead end, as the "stories" are still unfolding. In every case, most of the story is still missing. No mind, however great it is, can witness these differences in the decree of God among people and not be perplexed. The person is trying to know that which is completely unknown to everyone, except to God. Hence, looking into the secret/reason of Al-Qadar becomes a trap to the mind. This is why the Prophet ﷺ said:

وإذا ذُكِر القدر فأمسِكوا

When the Qadar (decree) of God is mentioned, hold your tongue in control.[70]

What could complicate the issue further is when the person tries to make sense out of what is happening to himself. Being in the middle of the problem, one's judgments are clouded to start with, and it becomes more difficult to see clearly. Difficulties and hardships shake one's belief more than anything else, and one mostly fails to see aspects of God's mercy during those moments. Thus, trying to understand "why" would only complicate the person's belief.

Such knowledge is inaccessible, and will always be so, regardless of what faith or perspective one is looking through. Expecting to know or understand before believing is the perfect example of "placing the carriage before the horse." *One cannot expect to understand what is happening in order to believe; quite the opposite, one would understand what is happening when he believes.* Once the person believes, everything falls in place.

A related question that people commonly ask is "if there is a God, why would He let so much misery happen?" The question can easily be answered in light of one's belief. The misery that we see is not the complete story. Our existence does not only include our life on this earth, but extends into the Hereafter. So there is still a continuation of what we experience in this life, into the afterlife, which is where the story is completed. People will receive their dues, and ultimate justice will be established, as emphasized repeatedly in the Qur'an.

وَنَضَعُ الْمَوَازِينَ الْقِسْطَ لِيَوْمِ الْقِيَامَةِ فَلَا تُظْلَمُ نَفْسٌ شَيْئًا وَإِن كَانَ مِثْقَالَ حَبَّةٍ مِّنْ خَرْدَلٍ أَتَيْنَا بِهَا وَكَفَى بِنَا حَاسِبِينَ

We shall set up scales of justice for the Day of Judgment, so that not a soul will be dealt with unjustly in the least, and if there be (no more than) the weight of a mustard seed, We will bring it (to account): and enough are We to take account.[71]

فَالْيَوْمَ لَا تُظْلَمُ نَفْسٌ شَيْئًا وَلَا تُجْزَوْنَ إِلاَّ مَا كُنتُمْ تَعْمَلُونَ

Then, on that Day, not a soul will be wronged in the least, and ye shall but be repaid the rewards of your past Deeds.[72]

People who were wronged in this world will be compensated over there; the transgressors will be punished.

But interestingly, even the logic behind the question itself is flawed. The one asking the question is starting with God, and then is questioning His Existence, based on His actions. To state the question differently, "There IS a God, He's allowing misery to happen; thus, I will deny His Existence." The one asking the question is *judging the actions of God, before even recognizing His Existence.*

Then, in the middle of all those unknowns of the Qadar (God's decrees), one must base his belief on something that is known, on a picture that is present and complete. The foundation for one's belief should be the whole universe, which testifies to the Existence and Greatness of God. Everything around us points to the Existence of an All-Wise, All-Great, All-Powerful, All-Generous God.

$$\text{سَنُرِيهِمْ آيَاتِنَا فِي الآفَاقِ وَفِي أَنفُسِهِمْ حَتَّى يَتَبَيَّنَ لَهُمْ أَنَّهُ الْحَقُّ}$$

Soon will We show them our Signs in the (furthest) regions (of the earth), and in their own souls, until it becomes manifest to them that this is the Truth.[73]

Thus, if one faces *one* source of doubt, he will still find *infinite* sources of belief, too many for a sound mind to ignore. And, when one does not see the "whys" of God's decree, he just *submits,* based on what he has learned of God's greatness through the whole world around him.

God knew that amidst life's hardship people will reach an edge in their belief and worship. Hence, He warned us to stick to that which we know through proofs and to stay away from the doubtful and incomprehensible, and that only He knows. We develop our knowledge of Him through His infinite creation, and then we reach a trusting point, where we submit to and accept His decree. We trust that everything He creates is ultimately for our good, even if we do not see it or understand it at the time.

$$\text{قُلِ اللَّهُمَّ مَالِكَ الْمُلْكِ تُؤْتِي الْمُلْكَ مَن تَشَاء وَتَنزِعُ الْمُلْكَ مِمَّن تَشَاء وَتُعِزُّ}$$
$$\text{مَن تَشَاء وَتُذِلُّ مَن تَشَاء بِيَدِكَ الْخَيْرُ إِنَّكَ عَلَىَ كُلِّ شَيْءٍ قَدِيرٌ}$$

Say: "O Allah. Lord of Power (And Rule), Thou givest power to whom Thou pleasest, and Thou strippest off power from whom Thou pleasest: Thou enduest with honour whom Thou pleasest, and Thou bringest low whom Thou pleasest: **In Thy hand is all good***. Verily, over all things Thou hast power.*[74]

When God gives power and when He strips off power, when He gives honor and when He brings low, it is all for a "good" reason.

4. Atheism

One problem that plagues the youth in particular is the erosion of belief, sometimes to the point of atheism. The problem seems to be no less serious in the other denominations. It may be caused by some doubt that had not been addressed, or, not uncommonly, by some other reasons not directly related to belief itself.

When examining the majority of books dealing with the Islamic Aqeedah, one easily notices that the focus is more on the belief in the Oneness of God (Tawheed) rather than the Existence of God. This might be due to the fact that atheism has been more of an exception through history. Deviations in belief were more related to the Oneness of God, rather than His Existence. But in the increasingly materialistic modern societies, the focus is shifting, and issues with belief are more related to the Existence of God.

In an extensive survey, Taunton[75] examined the factors affecting or associated with the decision to unbelieve for Christian college students. Though I have not undertaken such a study myself, I can make some anecdotal conclusions based on over 20 years of involvement in the religious scene, particularly with the youth in the Muslim community. Some of the factors he listed are specific to the Christian church and will be omitted, yet some of them appear universal:

1) Atheists are not completely foreign to the religion. Many of them were at some point connected to a mosque before deciding to unbelieve. Though not as strongly as indicated by Taunton, many of those atheists who had a prior connection to the religion express a dissatisfaction with the mosque and the religious people.

Some of them felt the mosque did not answer the pressing questions they had in their mind and may have even been met with reprimand for bringing up those questions. A rigid Imam or teacher may have contributed to push them further away from the religion. Additionally, some of them may have been pushed away from religion by some of the conflicts they commonly observe in Mosques and Islamic centers.

2) Emotions play a major role in embracing unbelief. People deciding to unbelieve often cite the suffering in the world and how it does not tell of a merciful God (see argument in the previous section). In the absence of a plausible explanation, these people find themselves veering toward disbelief. The "other side" becomes the default, simply because they did not find a satisfying explanation in the religion.

A rigid parent is also very commonly present in the picture. A protective parent who strongly cares to instill in his child the religious principles may resort to some very rigid measures. Instead of bringing the child into the religion, these measures sometimes push the son/daughter further away from the religion. When the son/daughter does not feel heard or seen by his/her overzealous parent, they rebel against the parent and everything the parent represents, including the parent's religion.

3) Teenage years are critical. Taunton believes the 14-17 years period to be decisive. It seems to be the age period when the impression is made, and a stand from the religion is taken, even if kept secretly.

4) The internet played an important role in their conversion to atheism. But with most people I believe the internet plays a catalyzing rather than a deciding factor. The internet provides a wealth of information on every topic, but people themselves seek specific selections. Someone who has some atheistic tendencies will find something that reinforces his beliefs, just as does someone with religious tendencies.

But, even for someone searching objectively, the internet can be tremendously confusing. Every belief and ideology is skillfully presented in the most convincing way.

In today's society, we are taught that everything is an exercise of eloquence and argumentation, and we even take pride in our ability to argue *for* and *against* the same issues. We can make both seem equally likely, even if one is much further away from the truth than the other. We have developed effective skills in conveying any ideology, even if erroneous. In an authentic hadeeth, the Prophet ﷺ said exactly that:

إن من البيان لسحراً ، أو إن بعض البيان لسحر

Indeed some people's expressive ability is magic.[76]

In the absence of a quality control, this "magical" presentation can be deceiving, and people exposed to it can be confused and even completely lost.

Presented as just anecdotal, my personal experience echoes the above. I have seriously and discretely considered unbelief during my teenage years (the 14 to17 period). Nothing about the way religion presented to me was appealing, and I would forsake my religious obligations whenever I could get away with it. I further held an unfounded grudge against religion and the religious people. What pushed me further in that direction were specifically factors mentioned above: 1) an "emotional factor," witnessing a devastating civil war, for many years, without seeing an end to it, and 2) the rigidity and close-mindedness of some of the "people of God." Exchange of religious ideas is not uncommonly shut off, and expression of different ideas is silenced and sanctioned.

What helped me out was meeting someone knowledgeable who listened, and to whom I shall forever be indebted. I had uncountable thoughts and questions, most of them irrational and flawed. I needed someone to listen and to recognize them, even if they were erroneous. Then I needed someone to help me correct them, which I was open to, after being acknowledged.

5. The Proof

The Existence of God is one of the most obvious realities. Every person feels It through his nature and is guided toward It through his Fitrah. It is the biggest reality, is self-evident, and requires no proof.

$$\text{قَالَتْ رُسُلُهُمْ أَفِي اللَّهِ شَكٌّ فَاطِرِ السَّمَاوَاتِ وَالْأَرْضِ}$$

Their apostles said: "Is there a doubt about Allah, The Creator of the heavens and the earth?"[77]

But sometimes that which is most obvious becomes overlooked. Additionally, the Western way of thinking has become prevalent in the whole world today, sharply criticizing all religions and challenging every belief, including the belief in God's Existence. It is even considered progressive to advance any form of challenge, even if one is challenging the most obvious. Thus, it would be beneficial to present some proofs to awaken the sleeping mind. God put signs of His Existence everywhere for people to see.

$$\text{سَنُرِيهِمْ آيَاتِنَا فِي الْآفَاقِ وَفِي أَنفُسِهِمْ حَتَّى يَتَبَيَّنَ لَهُمْ أَنَّهُ الْحَقُّ}$$

Soon will We show them our Signs in the (furthest) regions (of the earth), and in their own souls, until it becomes manifest to them that this is the Truth.[78]

Any intelligent mind contemplating this great universe, full of wisdom and beauty, will have unlimited evidence pointing to a supreme power that created and still controls the universe. Some have called it "the initial cause," "the initial mind," or the "initial mover." The Qur'an and earlier revealed books called it "Allah." He is the Great Creator whose existence can be easily and naturally recognized and accepted, but whose Self is beyond human perception and understanding.

$$\text{ذَلِكُمُ اللَّهُ رَبُّكُمْ لا إِلَهَ إِلاَّ هُوَ خَالِقُ كُلِّ شَيْءٍ فَاعْبُدُوهُ وَهُوَ عَلَى كُلِّ شَيْءٍ وَكِيلٌ}$$
$$\text{لاَّ تُدْرِكُهُ الْأَبْصَارُ وَهُوَ يُدْرِكُ الْأَبْصَارَ وَهُوَ اللَّطِيفُ الْخَبِيرُ}$$

That is Allah, your Lord! There is no god but He, the Creator of all things: then worship ye Him: and He hath power to dispose of all affairs.

No vision can grasp Him, but His grasp is over all vision: He is above all comprehension, yet is acquainted with all things.[79]

The Qur'an presented proofs for God's Existence in different ways. Sometimes it appeals to the sound, uncomplicated human Fitrah, which tells the human that she has a Creator, a Great Lord that oversees her.

فَأَقِمْ وَجْهَكَ لِلدِّينِ حَنِيفًا فِطْرَةَ اللَّهِ الَّتِي فَطَرَ النَّاسَ عَلَيْهَا لا تَبْدِيلَ لِخَلْقِ اللَّهِ ذَلِكَ الدِّينُ الْقَيِّمُ وَلَكِنَّ أَكْثَرَ النَّاسِ لا يَعْلَمُونَ

So set thou thy face steadily and truly to the Faith: Allah's handiwork according to the pattern (Fitrah) on which He has made mankind: no change (let there be) in the work (wrought) by Allah. That is the standard Religion: but most among mankind understand not.[80]

The Fitrah would be normally sufficient to guide a person to her Lord. In a hadeeth Qudsi, the Prophet ﷺ said that Allah says:

إِنِّي خَلَقْتُ عِبَادِي حُنَفَاء كُلَّهِمْ ، وَإِنَّهُمْ أَتَهُمُ الشَّيَاطِينُ فَاجْتَالَتْهُمْ عَنْ دِينِهِمْ

I have created all my servants on the pure state of Al-Fitrah, then the devils came to them and deviated them away from their religion.[81]

But the Fitrah fades under the pressure of worldly influence and preoccupation. However, it does not completely disappear; it re-surfaces under hardship.

هُوَ الَّذِي يُسَيِّرُكُمْ فِي الْبَرِّ وَالْبَحْرِ حَتَّى إِذَا كُنتُمْ فِي الْفُلْكِ وَجَرَيْنَ بِهِم بِرِيحٍ طَيِّبَةٍ وَفَرِحُواْ بِهَا جَاءَتْهَا رِيحٌ عَاصِفٌ وَجَاءَهُمُ الْمَوْجُ مِن كُلِّ مَكَانٍ وَظَنُّواْ أَنَّهُمْ أُحِيطَ بِهِمْ دَعَوُاْ اللَّهَ مُخْلِصِينَ لَهُ الدِّينَ لَئِنْ أَنجَيْتَنَا مِنْ هَذِهِ لَنَكُونَنَّ مِنَ الشَّاكِرِينَ

He it is Who enableth you to traverse through land and sea; so that ye even board ships;- they sail with them with a favourable wind, and they rejoice thereat; then comes a stormy wind and the waves come to them from all sides, and they think they are being overwhelmed: they cry unto Allah, sincerely offering (their) duty unto Him saying, "If thou dost deliver us from this, we shall truly show our gratitude!"[82]

Then, once the hardship is relieved, the person's doubts and confusions return; the Fitrah recoils, and the person asks again for a "proof."

And the proofs are many. Everyone finds a proof that satisfies him:[83]

- The simple person finds proofs for God's Existence, His Oneness and Greatness, that fit his level of thinking and education.

- The smart person examines deeper, with more specific and more profound proofs, and comes to the very same conclusion.

- The philosopher contemplates the reality using yet more profound and specific proofs that get into the deep secrets of things, and declares the same conclusion, the Existence of the Great Creator.

- The experimental researcher arrives through accurate, unbiased experimentation to a new proof connecting the material world to an initial wise cause, the Creator.

- The genius finds in his own field hundreds of proofs that make him submit to the idea of Existence of the Great Creator.

Sixty leading scientists, among them 24 Nobel laureates, were asked about God, the universe, and the human.[84] The great majority of the answers pointed in the same direction, the belief in the Existence of a Supreme Power or Great Creator behind this universe. Different scientists arrived to the very same conclusion differently:

"How can I exist without a creator? I am not aware of any compelling answer ever given." (Ulrich Becker[85])

"I believe that there is a God and that God brings structure to the universe on all levels from elementary particles to living beings to super-clusters of galaxies." (John Fornaess[86])

"I believe that God exists, but I do not have a personal understanding of what this means." (Vera Kistiakowsky[87])

"The existence of the universe requires me to conclude that God exists." (Robert Naumann[88])

"I believe that the universe was created and is sustained by some power that we call God." (John Russell[89])

"Now this sense of wonder leads most scientists to a Superior Being, a Superior Intelligence, the Lord of all Creation and Natural Law." (Abdus Salam[90])

"It seems to me that when confronted with the marvels of life and the universe, one must ask why and not just how. The only possible answers are religious.... I find a need for God in the universe and in my own life." (Arthur Schowlow[91])

"...nothing is more evident, more certain, than the evidence or reality of God." (Wolfgang Smith[92])

"We must admit that there exists an incomprehensible power and force with limitless foresight and knowledge that started the whole universe going in the first place." (Christian Afinsen[93])

"His existence is apparent to me in everything around me, especially in my work as a scientist." (Steven Bernasek[94])

"If I consider reality as I experience it, the primary experience I have is of my own existence as a unique self-conscious being which I believe is God-created." (John Eccles[95])

"To me, the concept of God is a logical outcome of the study of the immense universe that lies around us." (Thomas Emmel[96])

Indeed, as the Qur'an indicates:

شَهِدَ اللَّهُ أَنَّهُ لاَ إِلَهَ إِلاَّ هُوَ وَالْمَلاَئِكَةُ وَأُوْلُواْ الْعِلْمِ

There is no god but He: That is witnessed by Allah, His angels, and those endued with knowledge.[97]

In addressing proofs for God's Existence, the Qur'an appeals to an active searching process. It repeatedly calls for the person to contemplate this universe, what it calls "Tafakkur" (تَفَكُّر).

$$\text{إِنَّ فِي خَلْقِ السَّمَاوَاتِ وَالأَرْضِ وَاخْتِلافِ اللَّيْلِ وَالنَّهَارِ لآيَاتٍ لِأُولِي الأَلْبَابِ الَّذِينَ يَذْكُرُونَ اللَّهَ قِيَامًا وَقُعُودًا وَعَلَى جُنُوبِهِمْ وَيَتَفَكَّرُونَ فِي خَلْقِ السَّمَاوَاتِ وَالأَرْضِ رَبَّنَا مَا خَلَقْتَ هَذَا بَاطِلاً سُبْحَانَكَ فَقِنَا عَذَابَ النَّارِ}$$

Behold! In the creation of the heavens and the earth, and the alternation of night and day,- there are indeed Signs for men of understanding;

Men who celebrate the praises of Allah, standing, sitting, and lying down on their sides, and contemplate the (wonders of) creation in the heavens and the earth, (With the thought): "Our Lord! Not for naught Hast Thou created (all) this! Glory to Thee! Give us salvation from the penalty of the Fire.[98]

Concerning these verses, it was said that Ataa' narrated that once he went with Ubaid Bin Omair to the Mother of Believers A'esha and asked her from behind a veil about something amazing concerning the Prophet ﷺ. She replied, "Everything about him was amazing, but once he came to me during my night and his skin touched mine, then he stood up and told me *let me worship my Lord.* He performed his ablutions and prayed till his beard started dripping with tears, then prostrated till the floor got wet. He then laid down on his side until Bilal came to call for the Fajr Athan, and told him *O Messenger of Allah, what makes you cry, knowing that God forgave all your past and future sins? He replied How would I not cry and Allah revealed to me tonight '**Behold! In the creation of the heavens and the earth...**' Then he said Woe to the one who reads it and does not ponder about it."*[99]

As Abu Hamed Al-Ghazali[2,100] described it, "Tafakkur" is an active searching process, based on comparing and analyzing. It enriches the mind with knowledge and enlightens the heart with

---

[2]Abu Hamid Muhammad Bin Muhammad al-Ghazali (ابو حامد محمد محمد ابن محمد غزالي: 1058–1111), known mostly as Algazel in the West, was a Muslim theologian, jurist, philosopher, and mystic of Persian descent. One of the most influential Muslim figures in history, he is considered to be a Mujaddid or renewer of the faith, and is commonly referred to "Proof of Islam" (Hujjat Al-Islam).

66

that knowledge. It is not to be confused with day-dreaming, which is a passive process.

"Tafakkur" is the most efficient tool in changing a person, because it comes from within the person himself. For example, someone looking to favor the Afterlife over this life can do so either 1) by hearing someone stating that the Afterlife is greater than this one, then just believing him and following his directives, or 2) by contemplating this world, its nature and comparing it with the Afterlife, then coming himself to the conclusion that the Afterlife is more worthy. Clearly, the second way is far more effective, as the person is basing his conclusion on something he witnessed himself.

The same can be said concerning believing in God's Existence, His Greatness, His Mercy, His Wisdom,… Seeing aspects of them after an active contemplative process is the most solid way to establish a belief in them.

The Qur'anic verses calling for Tafakkur are too many to list; below are just a few.

إِنَّ فِي خَلْقِ السَّمَاوَاتِ وَالأَرْضِ وَاخْتِلافِ اللَّيْلِ وَالنَّهَارِ وَالْفُلْكِ الَّتِي تَجْرِي فِي الْبَحْرِ بِمَا يَنْفَعُ النَّاسَ وَمَا أَنْزَلَ اللَّهُ مِنَ السَّمَاء مِن مَّاء فَأَحْيَا بِهِ الأَرْضَ بَعْدَ مَوْتِهَا وَبَثَّ فِيهَا مِن كُلِّ دَابَّةٍ وَتَصْرِيفِ الرِّيَاحِ وَالسَّحَابِ الْمُسَخِّرِ بَيْنَ السَّمَاء وَالأَرْضِ لآيَاتٍ لِّقَوْمٍ يَعْقِلُونَ

Behold! in the creation of the heavens and the earth; in the alternation of the night and the day; in the sailing of the ships through the ocean for the profit of mankind; in the rain which Allah Sends down from the skies, and the life which He gives therewith to an earth that is dead; in the beasts of all kinds that He scatters through the earth; in the change of the winds, and the clouds which they Trail like their slaves between the sky and the earth;- (Here) indeed are Signs for a people that are wise.[101]

67

وَمِنْ آيَاتِهِ أَنْ خَلَقَكُم مِّن تُرَابٍ ثُمَّ إِذَا أَنتُم بَشَرٌ تَنتَشِرُونَ

وَمِنْ آيَاتِهِ أَنْ خَلَقَ لَكُم مِّنْ أَنفُسِكُمْ أَزْوَاجًا لِّتَسْكُنُوا إِلَيْهَا وَجَعَلَ بَيْنَكُم مَّوَدَّةً وَرَحْمَةً إِنَّ فِي ذَلِكَ لَآيَاتٍ لِّقَوْمٍ يَتَفَكَّرُونَ

وَمِنْ آيَاتِهِ خَلْقُ السَّمَاوَاتِ وَالْأَرْضِ وَاخْتِلَافُ أَلْسِنَتِكُمْ وَأَلْوَانِكُمْ إِنَّ فِي ذَلِكَ لَآيَاتٍ لِّلْعَالِمِينَ

وَمِنْ آيَاتِهِ مَنَامُكُم بِاللَّيْلِ وَالنَّهَارِ وَابْتِغَاؤُكُم مِّن فَضْلِهِ إِنَّ فِي ذَلِكَ لَآيَاتٍ لِّقَوْمٍ يَسْمَعُونَ

Among His Signs in this, that He created you from dust; and then,- behold, ye are men scattered (far and wide)!

And among His Signs is this, that He created for you mates from among yourselves, that ye may dwell in tranquility with them, and He has put love and mercy between your (hearts): verily in that are Signs for those who reflect.

And among His Signs is the creation of the heavens and the earth, and the variations in your languages and your colours: verily in that are Signs for those who know.

And among His Signs is the sleep that ye take by night and by day, and the quest that ye (make for livelihood) out of His Bounty: verily in that are signs for those who hearken.[102]

وَاللَّهُ أَنزَلَ مِنَ السَّمَاء مَاء فَأَحْيَا بِهِ الْأَرْضَ بَعْدَ مَوْتِهَا إِنَّ فِي ذَلِكَ لَآيَةً لِّقَوْمٍ يَسْمَعُونَ

وَإِنَّ لَكُمْ فِي الْأَنْعَامِ لَعِبْرَةً نُّسْقِيكُم مِّمَّا فِي بُطُونِهِ مِن بَيْنِ فَرْثٍ وَدَمٍ لَّبَنًا خَالِصًا سَائِغًا لِلشَّارِبِينَ

وَمِن ثَمَرَاتِ النَّخِيلِ وَالْأَعْنَابِ تَتَّخِذُونَ مِنْهُ سَكَرًا وَرِزْقًا حَسَنًا إِنَّ فِي ذَلِكَ لَآيَةً لِّقَوْمٍ يَعْقِلُونَ

And Allah sends down rain from the skies, and gives therewith life to the earth after its death: verily in this is a Sign for those who listen.

And verily in cattle (too) will ye find an instructive sign. From what is within their bodies between excretions and blood, We produce, for your drink, milk, pure and agreeable to those who drink it.

And from the fruit of the date-palm and the vine, ye get out wholesome drink and food: behold, in this also is a sign for those who are wise.[103]

أَفَلَا يَنظُرُونَ إِلَى الإِبِلِ كَيْفَ خُلِقَتْ
وَإِلَى السَّمَاء كَيْفَ رُفِعَتْ
وَإِلَى الْجِبَالِ كَيْفَ نُصِبَتْ
وَإِلَى الأَرْضِ كَيْفَ سُطِحَتْ

Do they not look at the Camels, how they are made?-
And at the Sky, how it is raised high?-
And at the Mountains, how they are fixed firm?-
And at the Earth, how it is spread out?[104]

فَلْيَنظُرِ الإِنسَانُ مِمَّ خُلِقَ

Now let man but think from what he is created![105]

فَلْيَنظُرِ الإِنسَانُ إِلَى طَعَامِهِ

Then let man look at his food, (and how We provide it)[106]

All the above verses call for a searching process. They give directions to look for the proof, which needs to be witnessed by the person herself. It is not sufficient to read about God's Greatness; one has to see It (aspects of It). Oftentimes, we read about something, then we think all the work is done. This might be true if one is only seeking information. But, if "change" is the goal, then books and lectures and information are only the beginning. Information gives directions and insights, but the change only comes from within after the person follows the directions. The person needs to witness the proof herself.

Muslim scholars presented various proofs of the Existence of God, some of them focusing on the same themes.[107,108,109,110]

A. Proof of the Perfect Design

As stated above, in the quest to learn about God, the Qur'an calls for an active searching process, using the simple law of causality, which points to an All-Wise Creator behind this universe. This perfectly designed universe must have a Creator to start it and a Sustainer to keep it running.

قَالَ رَبُّنَا الَّذِي أَعْطَى كُلَّ شَيْءٍ خَلْقَهُ ثُمَّ هَدَى

He (Moses) said (to the Pharaoh): "Our Lord is He Who gave to each (created) thing its form and nature, and further, gave (it) guidance."[III]

An infinite number of factors are simultaneously needed for the proper operation of every process in this intricate universe, from the physical to the biological. Randomness alone cannot explain the observed perfection.

Consider an exercise: If anyone were to mix the letters of the alphabet and randomly draw four letters, one at a time, what is the likelihood for the letters to form the word "LIFE"?

- For the first letter, the separate probability is 1/26 (26 is the initial number of letters)
- For the second letter, the probability is 1/25 (25 is the remaining number of letters after the first one was drawn)
- For the third letter, the probability is 1/24
- For the fourth letter, the probability is 1/23
- Then for "LIFE" to be randomly picked from a mixture of letters, the probability is $1/26 * 1/25 * 1/24 * 1/23 = 1$ *in 358,800*, or *0.00003 %.*

If we were to add just two more letters, "R" & "S", the probability for forming the word "LIFERS" goes down to *1 in 165,765,600*, or *0.00000006 %.*

If we were to increase the size of the word *and* we were to use a larger initial pool (e.g., if we add to the mix the Arabic alphabet, old Latin characters, Hindi characters, numerals…), then the probability to pick a specific word would be astronomically smaller.

Then if the number of options (the initial pool of numbers in the example above) is infinite, and the event we are looking for is very complex, the probability for it to happen randomly is zero, i.e., it would be an impossibility.

An example would be the perfect (and dynamic) alignments of the planets in the universe. Planets have masses; thus, they attract each other. But for planetary orbits not to be destabilized, and for one planet to not be drawn into another planet and collide with it, there needs to be a certain distance and an equal force pulling that planet in the opposite direction. Attractions in different directions need to be balanced against one another.

So what we have is:
- an infinite number of planets of different sizes,
- at different distances from one another,
- all attracting each other,
- while all are moving at different speeds,
- *yet* all are aligned in a perfectly balanced way.

One cannot even imagine the probability for that to happen, if it were to happen as a result of a totally random process. If any of those many and complex numbers was to be different, the system would not exist. The system has to be perfect in every way, with all the specific required numbers and elements being present at the same time. It cannot develop randomly out of a gradual process. Randomness cannot *lead* to it, and cannot explain it. There must have been a Designer who set all the correct parameters at the same time.

Interestingly, if one were to imagine lines of attraction forces between each planet and all the other planets, then a picture of an impossibly complex network or weave would be observed. This might be what this verse in Surat Al-Thariyaat refers to:

وَالسَّمَاء ذَاتِ الْحُبُكِ

By the Sky with (its) numerous Weaves.[112]

Then if we were to examine the possibility of life on earth, we find the whole cosmos giving every appearance of having been specifically and optimally tailored to that end.[113] Specific conditions, even those occurring far away from the planet, are required. Among them are the following:

- Specific placements of the planets (as mentioned in the example above).

- Nuclear furnaces in the interior of the stars in which hydrogen will be converted into the heavier elements essential for life.

- Specific proportion of the stars undergoing supernova explosions to release the key elements into interstellar space.

- Distribution and frequency of supernovae being not so frequent that planetary surfaces would be repeatedly bathed in lethal radiation, but not so infrequent that there would be no heavier atoms manufactured and gathered into the surface of newly formed planets.

- Specific distances from the sun.

- Specific distance from the moon.

- Specific orbit, with a specific speed following the orbit.

- Specific axis tilt, with a specific speed of rotation around the axis.

- Specific atmosphere composition.

- Specific combination of elements, both on the surface and deeper in the earth, with each element having specific characteristics.

Each of the above factors can then be broken further into many other factors, all of which are specifically needed in the exact way they are. For example, water has to have all of its present characteristics in order for it to support life (specific heat capacity, boiling point, density and expansion, surface tension, capillary action, etc.). If any of these parameters were to be different, life as we know it would not be supported.

Then life actually started. Though the start of life remains largely inexplicable, its existence is critically dependent on (but not caused by) the laws and constants of physics having the precise values they do. The parameters listed above are just examples. Everything had to be the way it is on earth itself: the carbon balance, the oxygen balance, the hydrological cycle, etc. If earth's nature had even a slightly different set of parameters, the world would be a very different place, and we would not be around to see it.[114] Furthermore, each constituent appears to be the *only* available candidate for its particular biological role, and is ideally fit in all its physical and chemical characteristics.

The picture becomes further complex when one examines the biological aspect. One just needs to glimpse the human's bodily functions to appreciate the intricacy of the system.

For instance, many illnesses and deficiencies can affect the human body; all of them are disturbances of normal functions, many of which we do not fully understand. The fact that the system can fail in so many ways, yet that it does mostly function properly, gives us a glance at the complexity of the human body. The same can be said about all animals, plants, and microorganisms.

Then the picture becomes yet more fascinating when one examines the interaction of the different organisms. Everything is linked in an intricately balanced ecosystem, where like strings in a spider web, every organism is linked to and affected by other organisms.

There is no end to the complexity and perfection, observed at every level in this universe. Everything points in the same direction, that behind all that magnificence must be some Great Designer, Creator, and Sustainer, Allah the Almighty.

صُنْعَ اللَّهِ الَّذِي أَتْقَنَ كُلَّ شَيْءٍ

(Such is) the artistry of Allah, who disposes of all things in perfect order[115]

B. Proof of the Necessary Being

The basic premise of this proof is that 1) everything in this world is either existent or non-existent, with no third option; and 2) the two states are mutually exclusive—something cannot be existent and non-existent at the same time and place.

The argument is simple yet may seem complicated; thus, I will present it in segments:

a. Which is the rule (vs. the exception), existence or non-existence?

"Non-existence" means the absence of the self and all the qualities of the non-existent; there would be no will, no thinking, no power...

Thus, something non-existent cannot bring itself into existence, for before existing it had no will, no power. That would be a contradiction.

Then, since "non-existence" cannot bring anything into existence, the rule must be "existence," meaning there has to be something, a cause, already existent that brings other things into existence.

This cause, whose existence is the "rule" (between mutually exclusive states: existence and non-existence), will be referred to as the "original cause."

If the existence of the "original cause" is the rule, then that existence does not require an explanation or a cause itself; otherwise, its existence would not be the rule. Explanations are needed only for that whose existence is not the rule.

Therefore:

1. The "rule" is existence

2. The "original cause," whose existence is the rule, does not require a cause or explanation; it is sufficient to say that its existence is the rule.

b. If existence of the "original cause" is the rule, then does the original cause have a beginning? And can it have an end, becoming eventually non-existent?

Firstly, the "original cause", whose existence is the rule, cannot have a temporal beginning; otherwise, its coming into existence (from non-existence) would require another cause. Then it would not be the "original" cause, and its existence would not be the rule.

Secondly, that whose existence is the rule cannot have an end or become non-existent.

"Non-existence" can only happen to one whose "non-existence" is the rule, when the factors required for its existence are removed. When these factors are gone, it goes back to its own rule, which is non-existence. Its existence is *contingent*, dependent on other factors.

But the "original cause," whose existence is the rule, cannot cease to exist, because there is no cause for its existence to start with. Its existence is *not contingent*.

Therefore:

3. That whose existence is the rule cannot have a start, and cannot become non-existent and have an end.

c. Then is "existence" the rule for everything we perceive in the world around us, or is it "non-existence"?

As humans, for example, we were non-existent, then we came into existence.

$$هَلْ أَتَى عَلَى الْإِنسَانِ حِينٌ مِّنَ الدَّهْرِ لَمْ يَكُن شَيْئًا مَّذْكُورًا$$

Has there not been over Man a long period of Time, when he was nothing - (not even) mentioned?[116]

$$أَوَلَا يَذْكُرُ الْإِنسَانُ أَنَّا خَلَقْنَاهُ مِن قَبْلُ وَلَمْ يَكُ شَيْئًا$$

But does not man call to mind that We created him before when he was nothing?[117]

Everything else we witness around also has a beginning, at different times: plants, animals, stars, and planets. Everything is continuously changing, whether we observe it or not, whether it is taking milliseconds or billions of years.

Thus, the rule for everything in our world is "non-existence," and their existence is contingent, requiring a cause.

Therefore:

4. For everything we perceive with our senses, their rule is "non-existence," and their coming into existence requires a cause.

d. Then the only explanation for the existence of the world is that *there was an Initial Cause that brought it into existence, a Cause Whose Existence Itself is the rule and is not contingent, a Cause Who is Existent by necessity.* It is the Great Creator of everything.

The Qur'an calls the human being to observe his own creation and think back about his cause of existence:

$$\text{أَمْ خُلِقُوا مِنْ غَيْرِ شَيْءٍ أَمْ هُمُ الْخَالِقُونَ}$$

Were they created of nothing, or were they themselves the creators?[118]
This verse asks "did they come from non-existence into existence without a Cause/Creator, or were they themselves the creators?" Both are impossibilities.

C. Proof of Change

A variation of the previous proof is the "proof of change." The world is in a constant state of change. From the subatomic to the galactic, everything is continuously changing: moving and transforming. Atoms bind into new molecules. Electricity transforms into heat. Solid water becomes vapor. The seeds grow into trees. Food is digested into nutrients. Young humans become old. Days become nights. An intensely hot summer becomes a cold winter. Planets move and change in conditions.

Change is sometimes quick, happening before our eyes. Change may also be slow, yet not less dramatic. A seed may take 20 years to grow into a large tree, gradually and unnoticeably. Yet the change is equally dramatic, should the same happen only in 20 seconds. Time is just relative.

But any change is a form of an incidental existence. It is easily observed with any change that involves transformation, e.g., a flower that becomes a fruit. Something was not, then became. This is true even for the simplest form of change, i.e., movement. Something was *not* in some location, then *became* in that new location.

76

But, for any observed change, a "changer" is needed. For anything to move or transform, there needs to be a cause for it to happen, and this is true for every step along the process of change. A changer/cause is needed.

Looking again at movement, no sound mind would accept that an inanimate object moves on its own without the action of a mover. It is the simple law of causality, observed throughout the universe. For every effect, there needs to be (a) cause(s). For example, a suitcase placed in a car cannot by itself move to an airplane. Someone, a changer, has to move it.

But what if there is a system of conveyors that would move the suitcases, without the need of "someone"? Then the conveyor system itself would be the "changer." Consequently, the question is now asked about the conveyor system: what is its changer? Who designed and created it, and who is operating it? For that system to come into existence, a changer has to bring it from non-existence into existence.

The questions then would keep rolling back until it reaches an "initial changer." But the initial changer itself does not change or need a changer itself; otherwise, it would not be the "initial changer." It is the independent source of all changes.

Changes that are much more complex than movement are constantly occurring. The more complex the change, the more complicated the changer. For a seed to germinate into a seedling and eventually grow into a tree, many changes are required at the physical, chemical, biochemical, cellular, and higher levels. Yet still, the deeper one looks into the change, the more enthralled he gets.

If one looks at a seed and compares it with the full grown tree, he gets fascinated.

If he looks deeper into the "changer" of that growth, the cellular processes, he gets more fascinated.

If he looks deeper into the changer of these cellular processes, the biochemical and genetic processes, he gets even more fascinated.

Looking yet deeper into the physical, chemical, atomic, and sub-atomic processes, the fascination still increases. Complex changes (seed to tree) start with some seemingly simple atomic processes: a perfect system that works perfectly at every level, to result at the end with beauty and ultimate perfection. Going deeper again, one eventually reaches an "initial all-wise changer" that itself does not change.

Everything is continuously changing, perishing, re-starting, disappearing, except for the "initial changer." Everything else needs a changer for it to exist, to change. The only real constant, the source of all changes that does not change is this "initial changer," Allah the Almighty!

$$كُلُّ مَنْ عَلَيْهَا فَانٍ$$
$$وَيَبْقَى وَجْهُ رَبِّكَ ذُو الْجَلَالِ وَالْإِكْرَامِ$$

All that is on earth will perish:
But will abide (forever) the Face of thy Lord,- full of Majesty,
Bounty and Honour.[119]

It is important to note that the above proofs tell us of the Existence of an Initial Cause, a Creator, despite us not being able to see that Cause. It is sufficient to recognize Its existence, without the need to look into Its nature.

As indicated earlier, under "Paths to Establish a Belief," our direct senses are limited in their perceptive abilities, and not being able to perceive something directly is not an indicator for its inexistence. Limitations are even more significant with the mind itself. Being unable to comprehend or imagine something does not mean it is inexistent. This is true even for some worldly things, where we accept their existence without understanding their nature or asking about their physical qualities. For example, we do not need to look into the nature of gravity in order to recognize its existence. Similarly, we recognize the connection between our mind and our physical body without having to understand its mechanism. The same can be said about many other aspects of our life, where we do believe in them without knowing their natures.

Limitations of the mind become more significant when trying to understand the nature of metaphysical realities. It is then sufficient to logically prove their existence, without having to conceive their nature.

Thus, our inability to conceive the Self of God should not be an obstacle to believing in His Existence. Very often, the attributes we ask to perceive reflect our anthropomorphic limitations, where we expect everything to be bound by the world that binds us. But the One who created the world cannot be bound by it or defined by the world's dimensions. It is sufficient to see signs of His Existence, and believing in It, despite being incapable of comprehending His Nature. He is, in His Self, beyond any human perception.

لاَّ تُدْرِكُهُ الأَبْصَارُ وَهُوَ يُدْرِكُ الأَبْصَارَ وَهُوَ اللَّطِيفُ الْخَبِيرُ

No vision can grasp Him, but His grasp is over all vision: He is above all comprehension, yet is acquainted with all things.[120]

We believe in Him as not resembling anything limited by our world.

لَيْسَ كَمِثْلِهِ شَيْءٌ وَهُوَ السَّمِيعُ الْبَصِيرُ

There is nothing whatever like unto Him, and He is the One that hears and sees (all things).[121]

6. Are Faith and Doubt Mutually Exclusive?

"Doubt isn't the opposite of faith; it is an element of faith."
— Paul Tillich[3]

Certainty in belief (i.e., certitude) is often over-emphasized

---

[3] Paul Johannes Tillich (1886 – 1965) was a German American Christian existentialist philosopher and highly influential theologian.

in our preaching heritage to the point that we came to consider anything short of certainty to be a "lack of faith." A slight doubt is commonly considered antithetical to belief. The problem gets more serious when people label *themselves* as disbelievers due to some doubt or satanic whispers that they experience.

For this reason, I believe the common perception, that *belief is either zero or 100%*, needs to be challenged. Though doubt diminishes belief, *belief and doubt are not mutually exclusive.* As Ibn Uthaimeen[122] stated (quoted in chapter "The Reality of Eeman"), belief increases or decreases with varying levels of conviction.

The Qur'anic verses and hadeeth that mention certainty, appear to describe the perfect state of Eeman and not its minimal requirements.

$$\text{إِنَّمَا الْمُؤْمِنُونَ الَّذِينَ آمَنُوا بِاللَّهِ وَرَسُولِهِ ثُمَّ لَمْ يَرْتَابُوا وَجَاهَدُوا بِأَمْوَالِهِمْ وَأَنفُسِهِمْ فِي سَبِيلِ اللَّهِ أُولَٰئِكَ هُمُ الصَّادِقُونَ}$$

Only those are Believers who have believed in Allah and His Messenger, **and** *have never since doubted, but have striven with their belongings and their persons in the Cause of Allah: Such are the sincere ones.*[123]

This verse is an example of verses describing the perfect belief. The belief in Allah and His Messenger would certainly be the minimal requirements of Eeman. But the "lack of doubt and the strivings with belongings and persons" would add to those minimal requirements, to bring Eeman up to perfection. The underlined conjunction "and" (ثُمَّ) here indicates an *addition*. No scholar has claimed that a lack of financial striving, for example, nullifies belief.

A similar conclusion may be derived from the following hadeeth, where the Prophet ﷺ said:

$$\text{أَشْهَدُ أَنْ لَا إِلٰهَ إِلَّا اللهُ وَأَنِّي رَسُولُ اللهِ ، لَا يَلْقَى اللهَ بِهِمَا عَبْدٌ غَيْرَ شَاكٍّ فِيهِمَا إِلَّا دَخَلَ الْجَنَّةَ}$$

"I bear witness that there is not god but Allah and that I am the messenger of Allah", anyone meeting Allah with it without having doubt in shall enter Al-Jannah.[124]

This hadeeth appears to refer to the perfect belief not diminished by doubt, and which guarantees a direct passage to paradise.

I am not discussing the *presence* of belief in the Oneness of Allah and in his Messenger. There is no dispute that the presence of such belief is a requirement for Eeman. But I am making the distinction between the *presence* of such belief and its *strength*. While its *presence* is required, its *strength* varies. A decrease in the strength of belief due to doubt does not eliminate the belief.

When the Sahabah went to the Prophet ﷺ complaining about what may sound like a doubt. The Prophet ﷺ acknowledged it as the "true faith", as in the hadeeth:

قال: جاء ناس من أصحاب النبي – صلى الله عليه وسلم – فسألوه: إنا نجد في أنفسنا ما يتعاظم أحدنا أن يتكلم به؟ قال: "وقد وجدتموه؟" قالوا: نعم قال : "ذاك صريح الإيمان"

Some of the Prophet's ﷺ companions came to him and asked him "Sometimes we find in our hearts some thoughts that we consider too serious for us to mention." He said "you found that?" They said "yes." He said "this is the true faith."[125]

The Prophet's ﷺ reply acknowledges that such thoughts are expected to be part of the "realistic faith." Thus, belief is not equivalent to certainty, though it may reach certainty in some moments. Those instances of certainty occur when belief reaches its perfect state, though not permanently. Permanent and constant certainty is not attained in this world. It will be achieved after death, when everything becomes witnessed directly. Thus, the verse:

وَاعْبُدْ رَبَّكَ حَتَّى يَأْتِيَكَ الْيَقِينُ

And worship thy Lord until there come unto thee the Hour of Certainty.[126]

The verse then would not mean "Worship your Lord until death comes to you" as commonly interpreted, but rather "Worship your Lord until your belief reaches certainty, which is brought forth by death." Until then, an occasional certainty, but mostly *plausibility*, would be expected.

Therefore, doubt is more related to human nature than to anything else. Doubt and certitude do not necessarily reflect the truthfulness of a religion or the validity of an idea. Doubt is a human trait, observed across religions, though to various degrees. Similarly certitude, or the illusion of it, may be attained for any idea and by anyone, without necessarily reflecting the validity of the idea. Philosopher Bertrand Russell[4] once wrote that "the stupid are cocksure while the intelligent are full of doubt."

One might get some valuable general insight into doubt by examining it through different religious perspectives. Some people who may be commonly considered firm believers may experience large patches of doubt in their life. One particular example worth citing is that of Mother Teresa[5]. Her posthumously published diaries[127] revealed, to the world's shock, that she was tormented by doubt herself. She regularly prayed for the Lord to "show Himself" and to dissipate the "terrible darkness within me, as if everything was dead." Someone of another faith than hers might easily argue that the cause of her doubt was the erroneous nature of her faith. I believe her doubt mostly reflects human nature; every human longs for certitude. But despite her complaints about doubt, her statements still reflect a general presence of belief. Her struggle is not unique.

Yet, even with the absence of certitude, some people just learn to live quite contentedly with doubt occasionally occurring

---

[4] Bertrand Arthur William Russell (1872 – 1970) was a British philosopher, logician, mathematician, historian, writer, social critic and political activist.
[5] Teresa of Calcutta, commonly known as Mother Teresa (1910 – 1997) was a Roman Catholic religious sister and missionary who lived most of her life in India. She was born in today's North Macedonia, with her family being of Albanian descent originating in Kosovo.

through their life. Salman Al-Audah[6,128] stated that the occasional doubt and confusion the believer experiences do not nullify his belief. The foundation of his belief remains much too strong to be neutralized by these thoughts. As Julia Baird[129] stated, no one can possibly hope to understand everything, and to have exhaustively researched all areas of uncertainty, in order to have certitude.

Therefore, an "uncertain belief" is NOT an "unbelief." Belief is not just an intellectual experience, based purely on rational factors. The heart, the rituals, and the habits are all parts of it. In the absence of certainty, plausibility might be a start and a good foundation for actions, which in turn feed back one's belief.

---

[6] Salman bin Fahd bin Abdullah Al-Audah (Arabic: سلمان بن فهد بن عبد الله العودة; born 1955 or 1956) is a Saudi Muslim scholar, member of the Board of Trustees of the International Union for Muslim Scholars. He is the founder and director of the website "Islam Today."

إِنَّمَا يَخْشَى اللَّهَ مِنْ عِبَادِهِ الْعُلَمَاءُ

*Those truly fear Allah, among His servants,
who have knowledge!*

BELIEF, ISLAM AND SCIENCE

A little bit of science leads one away from God, but much of science brings him back to Him
—Louis Pasteur[1]

One major source of doubt and confusion is the relationship between religion and science. This relationship has always been an important issue with all religions. People may learn something in one that seems to contradict the other, and often feel torn between the two. Particularly, it becomes a problem for someone young, who developed a strong belief within the familiarity of her own family environment, then gets exposed to some "new" scientific information in school that challenges that belief. Not being able to screen the credible out of the questionable, she becomes confused and doubtful. If the issues are not addressed, she might ultimately unbelieve, simply because she mistakenly thought that "no one may question science." Science is supposed to be the ultimate and unbiased standard. Should anything conflict with science, it is the former that falls.

Obviously, what is considered science includes a wide range of information, from the conclusive to the conjectural.

---

[1] Louis Pasteur (1822 – 1895) was a French chemist and microbiologist renowned for his discoveries of the principles of vaccination, microbial fermentation and pasteurization. Original quote: *Un peu de science éloigne de Dieu, mais beaucoup y ramène.*

The present chapter examines the relationship between the two. But, while some of the addressed issues are common to other religions, our discussion will focus on Islam. Some concerns (such as the age of the earth) raised by other religions are not relevant to Islam, either because they are addressed differently or just not addressed at all in Islam.

1. Definitions

In discussing the relationship between Islam and science, it is imperative first to define "religion" and "science," as some questions are answered in light of the definitions themselves.

A. Science

Webster's dictionary defines science as: "knowledge attained through study or practice," or "knowledge covering general truths of the operation of general laws, esp. as obtained and tested through scientific method [and] concerned with the *physical* world."[130] Restated differently, "science" refers to a system of acquiring knowledge, using observation and experimentation, to describe natural phenomena.

The Center for Science and Religion in Samford University defines science as the "acquisition of *reliable but not infallible* knowledge of the *real world*, including explanations of the phenomena."[131]

Less formally, the word science often describes any systematic field of study to gain knowledge. Thus, among others, there are:

1. Natural Sciences, that study the natural world;

2. Social Sciences, that study human behavior and society;

3. Various Islamic Sciences, that study various aspects of the religion.

Obviously, many scientists question such nomenclature, as they maintain that, for a body of knowledge to be considered a science, it must stand up to repeated testing by independent observers. As such, even social sciences may not be considered "sciences."

In this book we will use "science" to refer to natural and physical sciences.

B. Religion

Defining "religion" is far more complicated and problematic, for there are many "forms" of what people consider "religion." The Cambridge Encyclopedia of English Language[132] states: "... no single definition will suffice to encompass the varied sets of traditions, practices, and ideas which constitute different religions."

Etymologically, religion is derived from the Latin word *religare*, which means "to tie, to bind." Some definitions are given below:

The belief in a god or in a group of gods; an organized system of beliefs, ceremonies, and rules used to worship a god or a group of gods; an interest, a belief, or an activity that is very important to a person or group.[133]

A set of beliefs concerning the cause, nature, and purpose of the universe, especially when considered as the creation of a superhuman agency or agencies, usually involving devotional and ritual observances, and often containing a moral code governing the conduct of human affairs.[134]

An organized collection of beliefs, cultural systems, and world views that relate humanity to an order of existence.[135]

An organized system of belief that generally seeks to understand purpose, meaning, goals, and methods of spiritual things. These spiritual things can be God, people in relation to God, salvation, after life, purpose of life, order of the cosmos, etc.[136]

Sometimes used interchangeably with faith, a belief concerning the supernatural, sacred, or divine, and the practices and institutions associated with such belief. In its broadest sense some have defined it as the sum total of answers given to explain humankind's relationship with the universe. In the course of the development of religion, it has taken an almost infinite number of forms in various cultures and individuals.[137]

The term "religion," thus, refers both to the *personal* practices related to faith and to *group* rituals stemming from shared convictions.

Obviously, not all definitions are agreed upon. The terms *"spiritual"* and *"sacred"* add to the complexity of defining religion. Unless there are supreme beings or deities, most beliefs would not fall into this religious category. For that reason, *some* atheists have problems with definitions containing a deity or superhuman power. They do consider themselves followers of a religion/belief, though their belief denies any power other than man. As a result, some definitions have added a "philosophy of life," and as such would include all of Agnosticism, Atheism, conservative Christianity, Humanism, Islam, Judaism, liberal Christianity, Native American Spirituality, Wicca, and other Neopagan traditions as religions.[138]

Though we are only discussing Islam as a religion, I mentioned all of the above to show how one can easily pass a wrong conclusion about religion(s) through generalization. One might experience some issues with *his own religion*, then wrongly generalize his conclusion about *other religions*, which might be fundamentally different from his. A Hindu, for example, might find something in his religion that contradicts logic, then concludes that all religions do so, though the contradictions he found may be only present in his own religion. His conclusion (if correct) should be relevant only to his religion, and he cannot generalize, knowing how vastly different religions can be. This does happen commonly, mostly within the intellectual circles, especially in the West, where negative religious experiences are generalized and universal conclusions are made.

The opposite would also be true, from the Islamic perspective. Not uncommonly, a Muslim may pick issues raised regarding *other religions* and automatically assumes they would be issues in *his own religion*, which might be very different. For example, people of other faiths might raise basic theological or scientific questions concerning their faith. Then some Muslims automatically raise the same issues, though they may not even be issues in Islam. An example would be the anthropomorphic depiction of God in some other religions, or some of the scientific contradictions.

The focus of this study is obviously "Islam," as established by Prophet Muhammad ﷺ, and described in the Qur'an and the *authentic* tradition of the Prophet ﷺ. Unless indicated otherwise, "religion" will be used to refer to "Islam," and the two words will be used interchangeably.

2. Importance of Knowledge in Islam

"Islam liberates the mind, urges contemplation of the universe, honors science and scientists, and welcomes all that is good and beneficial to mankind." Those words, said by Hassan Al-Banna[2,139] summarize Islam's position regarding science and knowledge. The revelation started with a call to learn, "Iqra'" (read), and the Qur'an repeatedly praises the pursuit of useful knowledge and raises knowledgeable people into high status:

$$\text{قُلْ هَلْ يَسْتَوِي الَّذِينَ يَعْلَمُونَ وَالَّذِينَ لا يَعْلَمُونَ}$$

Are those equal, those who know and those who do not know?[140]

$$\text{إِنَّمَا يَخْشَى اللَّهَ مِنْ عِبَادِهِ الْعُلَمَاءُ}$$

Those truly fear Allah, among His Servants, who have knowledge.[141]

$$\text{يَرْفَعِ اللَّهُ الَّذِينَ آمَنُوا مِنكُمْ وَالَّذِينَ أُوتُوا الْعِلْمَ دَرَجَاتٍ}$$

---

[2] Hassan Ahmed Al-Banna (حسن أحمد البنا; 1906 – 1949) was an Egyptian school teacher and reformist, best known for founding the Muslim Brotherhood, one of the largest and most influential 20th century Muslim revivalist organizations.

Allah will rise up, to high ranks and degrees, those of you who believe and who have been granted Knowledge[142]

The pursuit of knowledge is not to be restricted to religious knowledge, but rather to extend to all spheres. The Qur'anic and Prophetic texts encourage people to seek any useful knowledge, whether religious or worldly.

One hadeeth might be particularly encouraging to medical researchers, where the Prophet ﷺ said:

$$\text{مَا أَنْزَلَ اللَّهُ دَاءً إلا أَنْزَلَ لَهُ شِفَاءً}$$

With every sickness that Allah sent down, He sent a cure.[143]

This hadeeth would certainly provide a great motivation for pursuing medical research. It is telling the believing researcher that she is looking for something that does indeed exist. The cure is actually present, and she is very likely to find it.

It is also important to note that the encouragement to seek knowledge is not restricted to men. A command given in the Qur'an or the Hadeeth normally addresses both genders. And, though marginalized in other societies, Muslim women historically played a significant role in the advancements of various sciences.[144]

History of Muslim Scientists

With a quick glance at Islamic history, one can easily notice that the glorious era of Islam was connected to significant scientific and technological advancements. Muslim scientists achieved unprecedented progress in various scientific fields, particularly during the Abbasid Caliphate and the Umayyad Caliphate in Al-Andalus[145,146]. Prominent scholars in various scientific fields were also firmly established in religious knowledge. Scientific revival was closely related to religious revival, unlike the way it was in the West, where the scientific revival happened when religion was separated from science.

Below are just a few examples of the significant contributions of some Muslim scientists:

Mathematics:

Muhammad Al-Khwarizmi (الخوارزمي)[3], considered the father of Algebra, had major contributions to mathematics, geography, astronomy, and cartography. It is noteworthy that he started his famous book "Completion and Balancing" (الجبر والمقابلة) with the *Basmallah.* The book was written following a request from the Caliph Al-Ma'mun, clearly in order to simplify resolving various matters of Islamic inheritance.

Medicine:

Sinan Bin Thabit Bin Qurrah (سنان بن ثابت)[4], ensured a control of the practice of medicine, where no one was allowed to practice medicine until going through an educational program, training and examination.

Abu Bakr Al-Razi, a.k.a. Rhazes (الرازي)[5,6], was another Muslim scientist and one of greatest scholars in medicine. Among his many contributions was the establishment of the *Bimaristan* (Persian for "hospital"), which were built all over the Muslim world, and run from Baghdad. These hospitals had reference libraries and different sections for men, women, and children. They even provided free medical care and financial compensation to workers during treatment periods. Al-Razi also emphasized the importance of mental health treatment In his famous book "Al-Haawi" (الحاوي).

---

[3] Muhammad Bin Mussa Al-Khwarizmi (Arabic: عبد الله محمد بن موسى الخوارزمي; 781-847), also known in the West as Algoritmi or Algaurizin, was a Persian mathematician, astronomer and geographer during the Abbasid Caliphate, and a scholar in the House of Wisdom in Baghdad.

[4] Sinan Bin Thabit Bin Qurrah (Arabic: 880-943 ;سنان بن ثابت بن قرة) was a physician, a physicist, and an astronomer from Harran, Syria. He moved to Baghdad where he converted to Islam after being a Sabian, and became the physician of the Abbassi Caliph Al-Muqtadir Billah.

[5] Abu Bakr Muhammad Bin Zakaria Al-Razi, (a.k.a. Rhazes, in Arabic: أبو بكر محمد بن يحيى بن زكريا الرازي; 923-864) was a Persian polymath, physician, alchemist and chemist, philosopher and important figure in the history of medicine.

[6] Not to be confused with another great scientist, Muhammad Bin Omar Bin Al-Hussein Al-Razi: known as Fakhreddin Al-Razi.

Al-Hussein Bin Abdillah Bin Sina, a.k.a. Avicenna (ابن سينا)[7], made tremendous contributions, writing almost 450 documents on a wide range of subjects, of which around 240 have survived. Of these, 150 books concentrated on philosophy and 40 on medicine. His most famous books were "The Book of Healing" (كتاب الشفاء), a vast philosophical and scientific encyclopedia; and "The Canon of Medicine" (القانون في الطب), an overview of all aspects of medicine, the standard medical text at many medieval universities until as late as 1650.

Physics:

Al-Hassan Bin Al-Haytham, a.k.a. Alhazen (ابن الهيثم)[8] is considered the father of optics. He discovered its basic laws and the way the camera obscura (in Arabic, "Qamarah" [القَمَرة]) as well as the human "camera-eye" work. He was one of the founders of the scientific method, which he used to verify the validity of information, even basic information, objectively.[147]

As with Al-Khawarizmi, most scientists had a strong basic and even advanced Islamic knowledge. A famous example was Abder-Rahman Ibn Khaldun (ابن خلدون)[9], regarded as one of the founding fathers of modern sociology, historiography and economics. He wrote the famous "Al-Muqaddimah" (المقدِّمة), a book on logic, Islamic jurisprudence, literature, words of

---

[7] Al-Hussein Bin Abdillah Bin Sina, a.k.a. Avicenna (Arabic: أبو علي الحسين ابن عبد الله ابن سينا; 980-1037) was a Persian physician and polymath who is regarded as one of the most significant thinkers and writers of the Islamic Golden Age.

[8] Abu Ali Al-Hassan Bin Al-Hassan Bin Al-Haytham, (Arabic: أبو علي، الحسن بن الحسن بن الهيثم; 965 – 1040), frequently referred to as Ibn Al-Haytham (Arabic: ابن الهيثم), known in the West as Alhazen, was an Arab polymath and philosopher who made significant contributions to the principles of optics, astronomy, mathematics, meteorology, visual perception and the scientific method.

[9] Abu Zayd Abdur-Rahman Bin Muhammad Bin Khaldūn Al-hadrami (Arabic: أبو زيد عبد الرحمن بن محمد بن خلدون الحضرمي; 1332 –1406), commonly known as Ibn Khaldoon (Arabic: ابن خلدون) was a Tunisian historiographer and historian, regarded to be among the founding fathers of modern sociology, and economics.

wisdom. For further information, the reader is referred to Souha Baayoun[148,149], Sigrid Hunke[150], and Al-Skaaf & Matraji[151], among the many available references.

One might thus conclude that the credit of the current development and technical advancement of the West is owed in part to the Muslim scientists. It is the result of the accumulation of knowledge and experience through history, in which Muslim scientists played a major role. The work of earlier scholars, both Muslims and non-Muslims, was essential, but the Muslim scholars in particular were able to take from those before them and add to it, qualitatively and quantitatively. Orientalist George Sarton[10] stated that the Arabs were the best teachers in the world, and, if they had not conveyed to the West the treasures of the Greek wisdom, the advancement of civilization would have stopped for several centuries. If it wasn't for Ibn Rushd (a.k.a. Averroes),[11] no one would have understood the work of Aristotle, whose teachings were essential for the development of universities in the West. The presence of Ibn Al-Haytham and Jabir Bin Hayyan and their likes was essential for the appearance of Galileo[12] and Newton[13]. But it

---

[10] George Sarton (1884–1956) was a Belgian-American chemist and historian who is considered the founder of the discipline of history of science.

[11] Abu al-Walid Muhammad ibn Ahmad ibn Rushd (أبو الوليد محمد بن أحمد بن رشد؛ 1198–1126), commonly known as Ibn Rushd (Arabic: ابن رشد) and referred to in the West as Averroes or The Commentator, was a Berber-Andalusian polymath of the medieval Islamic world. He wrote extensively on logic, Aristotelian and Islamic philosophy, theology, the Mālikī school of Islamic jurisprudence, psychology, political theory, Andalusian classical music, geography, mathematics, and the medieval sciences—including medicine, astronomy, physics, and celestial mechanics. A 13th-century philosophical movement based on his work became known as Averroism.

[12] Galileo Galilee (1564 – 1642), often known mononymously as Galileo, was an Italian physicist, mathematician, engineer, astronomer, and philosopher who played a major role in the scientific revolution during the Renaissance.

[13] Isaac Newton (1642 – 1727) was an English physicist and mathematician, widely recognized as one of the most influential scientists of all time, most famous for his law of gravitation. He was instrumental in the scientific revolution of the 17th century.

should be emphasized again that while those Western scientists thrived as a result of the separation of science and religion, it was the opposite with the Muslim scientists.

3. Relationship between Islam and Science

Knowledge in different fields is of different natures. It can be biological, physical, social, artistic, psychological, or metaphysical. Being of different natures, obviously, different knowledge is pursued differently, through different means and using different tools. One cannot use physical tools to *directly* measure social issues, or use artistic abilities to *directly* develop medical cures.

Still, some of the various disciplines might sometimes overlap, covering similar areas, though from different perspectives. For instance, though religion and science have different scopes, they might still overlap in some areas. Islam might address some issues of scientific nature, for the following reasons:

1. To provide some worldly guidance. Some Qur'anic verses and hadeeths provide some specific instructions that people need in their daily lives, which can be regarded scientific in nature:

وَكُلُواْ وَاشْرَبُواْ وَلاَ تُسْرِفُواْ إِنَّهُ لاَ يُحِبُّ الْمُسْرِفِينَ

2. *And eat and drink: But waste not by excess, for Allah loveth not the wasters[152]*

كل مسكر خمر ، وكل مسكر حرام ، ومن شرب الخمر في الدنيا فمات وهو يدمنها لم يتب لم يشربها في الآخرة

3. *Anything that leads to drunkenness is Khamr (wine), and all wine is prohibited. And whoever drinks wine in this life and dies while still a wine-drinker before repenting, will not drink it in the Hereafter.[153]*

Similarly, there are instructions concerning cleanliness, purity, and health, as with the whole area of Prophetic medicine.[154]

To confirm the divine nature of the Qur'anic text, and that it could not have been the words of a human. Several verses describe some scientific facts that were not discovered until much later. An example is the fact that the child's gender is determined by the father (giving an 'X' or a 'Y' sex-chromosome), not the mother (giving only an 'X' sex-chromosome). This is clearly stated in Surat Al-Qiyamah:

$$...أَلَمْ يَكُ نُطْفَةً مِّن مَّنِيٍّ يُمْنَى$$
$$فَجَعَلَ مِنْهُ الزَّوْجَيْنِ الذَّكَرَ وَالأُنثَى$$

Was he (the human) not a drop of sperm emitted?...
And of it (the sperm) He made two sexes, male and female[155].

The reader is referred further to the book by Maurice Bucaille[156] and various writings by Zaghloul El-Naggar[14].

4. To emphasize the cohesiveness of the various spheres of life, scientific, religious, and otherwise. Life is a continuum where all are parts of the same "whole," even if they are different aspects. As such, all can work toward the same general goals without conflicts. Islam emphasizes that concept though its comprehensive nature as in the verse:

$$قُلْ إِنَّ صَلاتِي وَنُسُكِي وَمَحْيَايَ وَمَمَاتِي لِلَّهِ رَبِّ الْعَالَمِينَ$$

5. *Say: "Truly, my prayer and my service of sacrifice, my life and my death, are (all) for Allah, the Cherisher of the Worlds.*[157]

Separations of various fields of study are purely academic and do not reflect in the oneness of life.

But with the overlap between Islam and science, Islam still

---

[14] Zaghloul El Naggar زغلول النجار, born 1933 is an Egyptian geologist, Muslim scholar, and Muslim author. The main theme of El-Naggar's books has been science in Qur'an. He is the Chairman of Committee of Scientific Notions in the Qur'an, Supreme Council of Islamic Affairs, Cairo, Egypt.

recognizes that science covers certain areas that religion does not concern itself with. This is true for the details of physical life, medicine, various sciences, some aspects of history, etc. Thus, Islam does not directly deal with all of the details of all of the spheres of life. For instance, in the field of "Fundamentals of Islamic Jurisprudence" (أصول الفقه), one source of legislation is the "Unrestricted Matters of Benefit" (المصالح المُرسَلة), which are matters of benefit not specified by the religion, and where various knowledge, not the religion, are used to determine their validity.

A Muslim does hold the basic belief that the Qur'an is a *complete book*, as described in Surat Al-Nahl:

وَنَزَّلْنَا عَلَيْكَ الْكِتَابَ تِبْيَانًا لِّكُلِّ شَيْءٍ وَهُدًى وَرَحْمَةً وَبُشْرَى لِلْمُسْلِمِينَ

And We have sent down to thee the Book explaining all things, a Guide, a Mercy, and Glad Tidings to Muslims.[158]

Nevertheless, the Qur'an is specifically a *complete book of guidance*. It guides humans to the fundamental truth and to their purpose; it shows them the way to live that purpose, the way to happiness. It addresses areas that science does not address or even understand, such as worship and metaphysical matters. But the Qur'an does not claim to be a complete book of everything. In the area of history, for example, the Qur'an says in Surat Ghafir that some historical information is omitted, for it is not essential for the *guidance* the Qur'an is meant to provide:

وَلَقَدْ أَرْسَلْنَا رُسُلًا مِّن قَبْلِكَ مِنْهُم مَّن قَصَصْنَا عَلَيْكَ وَمِنْهُم مَّن لَّمْ نَقْصُصْ عَلَيْكَ

We did aforetime send messengers before thee: of them there are some whose story We have related to thee, and some whose story We have not related to thee.[159]

It also commands the believers to consult with the experts in various fields should they seek answers:

فَاسْأَلُواْ أَهْلَ الذِّكْرِ إِن كُنتُمْ لاَ تَعْلَمُونَ

If ye know not, ask of those who possess the Message/knowledge[160]

Thus, Islam and the Qur'an are not meant to answer all sorts of questions, some of which are left to the worldly sciences. Only questions related to guidance are addressed. That is why the Islamic revelation omits much biological information, for example. Interestingly, I have been asked many times about what Islam says concerning dinosaurs. The answer is "nothing," because the Qur'an was not meant to be a book of natural history. Similarly, Islam does not say anything about the age of the universe, because it is not meant to be a book of astronomy.

But, while the Qur'an can claim to be a complete book of guidance, science cannot make the same claim of its own. No one can, nor would ever, claim that science knows everything it deals with. While "guidance knowledge" is complete in the Qur'an, "scientific knowledge" is much far from being complete. Despite all scientific advancements, what we ignore remains far more than what we know.

For example, one fundamental question is *whether the human is molecular/neuronal.* Is the human a simple assemblage of molecules, where every cerebral function is traceable to a molecule or some physical component?

This question was thought to have been answered since the publication of Darwin's "Origin of Species" in 1859. Scientists have traced various functions to some molecules and different areas in the brain… except ONE, the major one: *spirit and consciousness* (Grousset[161]). Interestingly, the Qur'an mentions the human lack of knowledge specifically regarding that human component:

وَيَسْأَلُونَكَ عَنِ الرُّوحِ قُلِ الرُّوحُ مِنْ أَمْرِ رَبِّي وَمَا أُوتِيتُم مِّن الْعِلْمِ إِلاَّ قَلِيلاً

They ask thee concerning the Spirit. Say: "The Spirit (cometh) by command of my Lord: of knowledge it is only a little that is communicated to you, (O men!)"[162]

Further troubling for those claiming the molecular nature of the human: 10% of victims of near-death experience comas that were brought back to life after the brain stops could recall certain perceptions that are visual, auditive, or intellectual (Grousset 91[163])! Thus, it seems that:

1) Consciousness (or is that the spirit?) can still function without the brain and without molecular support.

2) Biological information is not only generated/transmitted through molecules, but also possibly at some platform or through a network that we don't know anything about yet.

3) We cannot pretend to know the whole (i.e., the human being) since we do not even know the composition of its parts.

This is not intended to discredit science and scientific findings. Life's physical and psychological quality has tremendously improved as a result of scientific discoveries in various fields. But science cannot, and must not, claim to know everything with conclusiveness. And *if* science has answered some *"how"* technical questions, it remains incapable of answering the *"why"* questions, which are to be answered only by religion.

But there is still an area of overlap where the relationship between science and religion must be addressed.

Overlap between Science and Religion

While examining the overlap and interaction between science and religion, one basic fact that is crucial to remember is that they are simply different paths to attain knowledge. While *science is knowledge attained through* **experimentation and observation** (which are strictly human), *religion is knowledge obtained through* **revelation**. Thus, *if the religion is true and science is correct, they should lead to the same conclusions.* Contradiction would only occur if one or both of them are flawed and thus erroneous! The Qur'an states that the revelation declares the oneness of God, a truth also attained by the people of knowledge:

شَهِدَ اللَّهُ أَنَّهُ لَا إِلَهَ إِلاَّ هُوَ وَالْمَلاَئِكَةُ وَأُوْلُواْ الْعِلْمِ

There is no god but He: That is the witness of Allah, His angels, and those endued with knowledge[164]

If the truth is one, accurate ways of seeking that truth should NOT lead to conflicting results. For that reason, it is important to determine the credibility of science and religion in leading to the truth by examining the level of certainty of knowledge derived from science and from religion.

As discussed in the second chapter, human means to acquire knowledge have limitations, due to perceptive and/or analytical limitations. With science being based on observation and experimentation, some of those limitations may introduce an element of error into science, despite the scientific process. Therefore, **scientific knowledge** can be:

1) Conclusive: This refers to knowledge that is certain, irrefutably verified, such as scientific facts and physical laws.

2) Probabilistic: This refers to knowledge that is uncertain. It includes unproven theories and unverified information, and it comprises the majority of scientific research. Probabilistic scientific knowledge includes a wide range of scientific information that varies in its credibility, some being highly probable, others being merely conjectural.

Though science uses observation as a tool, still, not all scientific findings are certain. This does not contradict what was mentioned in the previous chapter, concerning direct observation/perception leading to Certitude of Sight (عين اليقين). An observation or direct perception does lead to certainty, but only concerning that which is observed itself. The certainty does not necessarily extend to *extrapolations* and *interpretations* of what is observed. Other factors may affect our interpretations, and the conclusion would not be as certain as the initial observation. For example, one may observe the current weather in a particular location, and her knowledge concerning this weather at that specific time in that particular area is conclusive and certain. But if she were to use that observation to predict a future weather or guess the weather in another area, her conclusion/knowledge would obviously not be as certain.

<u>**Religious knowledge**</u> itself can also be:

Conclusive: This describes knowledge that is at the same time conclusive in its authenticity *and* deterministic in its meaning. Idiomatically, it refers to a text that is Qat'ee Al-Thuboot (قطعيّ الثُّبوت) and Qat'ee Al-Dalalah (قطعيّ الدّلالة). It includes Qur'anic verses and Mutawater hadeeths (أحاديث متواترة), both of which are conclusively authentic, having clear and well-defined meanings.

Probabilistic: This refers to information that is uncertain in authenticity *and/or* in meaning. Idiomatically, it describes a text that is Thanny Al-Thuboot (ظنّي الثُّبوت) and/or Thanny Al-Dalalah (ظنّي الدّلالة). This include:

A) Qur'anic verses (though conclusively authentic) that have unclear meaning,

B) Hadeeths that are conclusively authentic, but have unclear meaning,

C) Hadeeths that are unauthentic, but have clear meaning; and at the lowest level,

D) Hadeeths that are both unauthentic and have unclear meaning.

Contradictions between scientific knowledge and religious knowledge (i.e., knowledge attained through experimentation and observation, and knowledge attained through revelation) occur in the following situations:[165]

1) That which is attributed to science has not reached the level of certainty; it is probabilistic and still needs further research, examination, and verification.

2) That which is attributed to the religion has not reached the level of conclusiveness in its authenticity (Thannee Al-Thuboot). Islamic narrations might be unauthentic, weak, or even fabricated, and as such may not be used as sources of Islamic knowledge.

3) Our understanding of the religious text is erroneous or not definitive (the text being Thanni Al-Dalalah to start with), even if the text itself is conclusive in its authenticity. The Qur'an itself

indicates that some of the verses have clear deterministic meaning while other have unclear allegorical meaning:

هُوَ الَّذِيَ أَنزَلَ عَلَيْكَ الْكِتَابَ مِنْهُ آيَاتٌ مُّحْكَمَاتٌ هُنَّ أُمُّ الْكِتَابِ وَأُخَرُ مُتَشَابِهَاتٌ

4) *He it is Who has sent down to thee the Book: In it are verses basic or fundamental (of established meaning); they are the foundation of the Book: others are allegorical.*[166]

Hassan Al-Banna summed up the relationship in his statement: *"religion and science (intellectual knowledge) may address different spheres, but will never conflict when both are conclusive. It is impossible for an established scientific fact to contradict an established authentic Islamic principle. If one of them is uncertain, then it should be reinterpreted in light of the conclusive other. If both are uncertain, then the uncertain Islamic principle should be given precedence over the uncertain scientific notion until the latter is proven."*[167]

Therefore, one may interpret a scientific finding in light of an Islamic text, if the former is uncertain and the latter is certain. Similarly, one may interpret an uncertain Islamic text in a way for it to fit the reality or a certain scientific finding. Al-Midani[168] strongly believes that we can *give a specific meaning to an otherwise general principle or Islamic text* (تخصيص العُموم وتأويل النَّص) in light of a scientific finding, so long as there is stronger support for the scientific finding. It may be even essential when the scientific is certain.

But strong care must be taken:

1) Not to interpret the religious text in light of an uncertain scientific finding. It is not uncommon to find some motivated religious person, stretching the meaning of some Qur'anic verses so as to fit some unverified scientific findings. Then should that particular scientific finding prove to be false, or even questioned, criticism is drawn to the Islamic text itself. But, obviously, what should be discredited in such a situation is our own understanding and interpretation of the religious text, and not the text itself.

103

2) Not to go into a post-modernist interpretation of the Ghaib (that which is unwitnessed and/or metaphysical). Some may attempt to *subjectively* explain certain matters of Ghaib, mentioned in the Qur'an or the hadeeth according to the laws of nature, in ways that seem logical to them. We see it happening with some writers, such as with Muhammad Asad's commentary on the Qur'an,[169] where certain stories and beings were interpreted as such (e.g., describing the birds in Surat Al-Fil as birds or insects carrying some epidemic).

It should be remembered that the metaphysical Ghaib (Jins, Angels, grave, etc.) is part of a different world, governed by different laws; thus, it cannot be interpreted in light of our world's laws of nature. Hence, we accept whatever is mentioned of that category without attempting to interpret its nature.

Similarly, when prophetic miracles are described (the snake of Mussa, the Whale of Yunus, etc.), they are mostly meant to prove the prophethood of the prophets, by presenting to people something challenging that defies the laws of causality. Trying to explain a miracle by extremely stretching the meaning of the words so as to fit the normal laws of nature defeats the very purpose of a miracle.

4. Evolution

When the relationship between science and religion is discussed, no topic can claim the central spot stronger than the Darwinian theory of evolution. Along with the Copernican heliocentrism[15], it was possibly the scientific theory that had the most profound

---

[15] Copernican heliocentrism is the name given to the astronomical model developed by Nicolaus Copernicus and published in 1543. It positioned the Sun near the center of the Universe, motionless, with Earth and the other planets rotating around it. The Copernican model departed from the Ptolemaic system that prevailed in Western culture for centuries, placing Earth at the center of the Universe, and is often regarded as the launching point to modern astronomy and the Scientific Revolution.

impact in this relationship (Ortoli[170]). The theory by its nature has many religious and philosophical implications, making it of concern not only to the scientists but also to the philosophers and even to the laypersons.

Still today, discussions around the Darwinian theory tend to be heated, with zealots on both sides making all sorts of accusations. Proponents of the theory are automatically categorized as atheists, while opponents are accused of imposing a religious perspective.

While the discussions have mostly involved people of Christian faith, some Muslims are increasingly taking part in them. Yet I do maintain that most of the issues raised in those discussions are not relevant to Islam, as the contradictions found in other religions with the theory are not found in Islam in the first place.

Regardless, the wide acceptance of this theory led to the elimination of God from nature, contributing significantly to the secularization of modern Western society.

A. Charles Darwin

Charles Darwin was born February 12, 1809, in a well-to-do middle class family in Shrewsbury, England He started as an amateur naturalist, a beetle collector. He flirted with the idea of becoming a physician, then a preacher in his early 20s, before shipping out of Davenport with the HMS *Beagle*, December 1831.

The observations he made during this 5-years trip were the basis of his views and the foundation of his theory.

Two years after he returned from the *Beagle*, shortly before he turned 30, Darwin married his first cousin Emma, with whom he had 10 children. He became very sick afterward, suffering from a variety of crippling stomach ailments and chronic headaches. He became intensely depressed and isolated. He stayed in bad health until he was 60. Bowlby[171] strongly believes that Darwin was hypochondriacal, i.e., his illnesses were psychosomatic, with no underlying physical cause.

He published his "The Origin of Species" when he was 50,

in 1859, and he defended his theory at the Oxford evolutionary debate in 1860.

But, despite his impact on the philosophical and religious landscape, Darwin considered himself as agnostic rather than atheist. He wrestled with religious doubts most of his life. Most troubling to him was the problem of evil. He had problems comprehending how a benevolent, omnipotent God could permit such suffering in the world He created.[16] What exacerbated his despair further was the death of his daughter Annie at the age of 10.

From the age of 60 until his death at 73, he was generally better in health. He died in Downe, England, in 1882, and was buried in Westminster Abbey.

More information on Darwin's biography may be found in many references[172].

B. The Theory

While on his voyage onboard the *Beagle*, Darwin had with him a book that was to have an increasingly powerful influence on his thinking as the journey progressed. It was the recently published Charles Lyell's famous "Principles of Geology"[173], in which Lyell described a world shaped by gradual changes across time. Lyell contended that the presently occurring geological processes were the same as the ones that occurred through history. As such, geological history can be observed in the present. Moreover, all occurring changes are gradual. This last idea colored everything Darwin observed on his trip. He noticed that different geographical regions were often populated by distinct yet closely related species. This clearly presented a challenge to the doctrine of fixity of species, which was prevalent at the time.

Then, there was Darwin's famous visit to the Galapagos Islands,

---

[16] This question was addressed in the previous chapter, under "Causes of Doubt."

a small Pacific archipelago, 600 miles east of Ecuador. Among other animals inhabiting the islands, there were 14 species of finches (later, famously known as the "Darwin finches"). They had great morphological and behavioral differences yet were closely related, belonging to the same subfamily of finches. Some of them could even be arranged into a morphological sequence. Thus, the idea that they are commonly related to an original ancestral species was compelling. Also, he observed that the fauna on the closest mainland was generally related to the one on the Galapagos, with sister species. Darwin's ideas started taking shape, and the belief in a gradual change started to develop. Hence was sown the seed of what became his theory, which he published 20 years later.

By the end of his trip on the *Beagle*, Darwin had already formed his ideas, that, contrary to popular belief, life is not static, and species change and evolve. But he did not understand the mechanism yet. He partially identified it 2 years later as natural selection, after reading economist Thomas Malthus[17] on the competition for resources among humans. Malthus argued that competition was caused by overpopulation, due to the fact that humans grow according to a geometric progression while resources grow according to an arithmetic progression. Darwin concluded that natural selection explains everything, that species with an adaptive advantage thrive and crowd out the rest. But, on his return, he kept his ideas under wraps for two decades.

He understood the implications of his theory, marginalizing

---

[17] Thomas Robert Malthus (1766–1834) was a British scholar and Anglican cleric best known for his pioneering ideas in economics and population theory. In his influential work *An Essay on the Principle of Population*, he argued that human populations tend to grow faster than the resources needed to sustain them. As a result, he warned that without natural checks such as famine and disease, societies would face inevitable hardship—a scenario now referred to as a Malthusian catastrophe. Malthus was skeptical of utopian visions of progress, emphasizing that population growth would always outstrip the Earth's ability to provide enough food and resources. As he famously put it, the growth of the population is "indefinitely greater than the power in the earth to produce subsistence for man."

the human being, regarded so far as the center of the world, into an aimless and mechanical existence. He wrote that publishing his theory was like "confessing a murder." Still, Darwin seemed genuinely trying to find answers and was not himself conclusive about his ideas.[174]

Darwin's famous model that describes the origin of species in nature can be summarized as follows:

1) The world is not immutable; species are under continuous modification. They are evolving.

2) This evolution is gradual. It does not happen by leaps, but through small variations that confer to the species some advantages or disadvantages.

3) Evolution is the result of natural selection. This happens through two components: the appearance of a great variation within the same species, followed by a selection, among this mix, of organisms that are more adapted to their environment. In other words, all living beings are involved in a continuous struggle for survival, and out of that struggle only the fittest make it. These fitter individuals have a marginal advantage, which allows them to dominate. Then, gradually and continuously, the accumulation of these advantageous traits results in a new living species, very distinct from its distant ancestors.

4) All living organisms have a common origin. In other words, there is a continuity of the living world, with all living forms, animal and vegetal, deriving from one another through filiation.

Obviously, a description of the mechanism was missing in the model above, namely 1) the source of variation within the species, and 2) a mechanism by which the gradual changes are passed to the following generations. The "gene" and the laws of heredity were not discovered yet.

Obviously, with the discovery of "genes," Darwin's theory was enriched and amended. The new and most common form within the Darwinian tradition became the synthetic theory of evolution or Neo-Darwinism, which integrates genetics into original Darwinism. The concept behind Neo-Darwinism, however, remains essentially the same: diversity, followed by natural selection.

Examining the main component of the theory, i.e., natural selection, Jonathan Wells[175] remarked that while natural selection certainly ensures the survival of the fittest, it does not *create* the fittest. It can never be the source of all adaptation and diversity that we see, as Darwin claimed. The fittest is selected for, but natural selection says nothing as to where the *fittest* initially came from.

Natural selection does lead to local adaptation, but also to the loss of diversity eventually. A selective pressure favors one particular form within a diverse population. As a result, it eliminates other forms, resulting in their local extinction.

Therefore, natural selection can only work with the already existent genetic information, selecting among the already existent genotypes. It never creates information, but rather it causes the loss of information, i.e., the loss of diversity.

Then the big question and large mystery is, "Where did that complexity of life come from?" What is the origin of the diversity, on which natural selection operates? The given answer was "random mutation." According to evolutionists, we have beneficial random mutations, followed by natural selection. But a major problem with that model is that mutations that take place on DNA are never constructive (Giuseppe Sermonti[176]). They can be locally advantageous but can never be constructive, because they always bring the loss of something, a function. Whatever mutated was, before the mutation, coding for a certain function, which is now lost. Therefore, examination at the molecular level indicates that information is lost by mutation, not gained.

It should also be noted that random mutations create aberrations far more readily than constructive characteristics. Then, mathematically, natural selection reduces diversity much faster than would be re-established by mutation.

According to Denton[177], the Darwinian model might explain in certain (but not all) cases some secondary changes between two closely related species. However, it does not explain the immense gaps between major divisions and the main essential phases of the evolution (apparition of new organs, new strata [birds, vertebrates, etc.]). Jean Drost[178] maintains that Darwin's theory has an abusive extrapolation. The mechanisms that explain how a small evolutive distance can be traversed are not the same mechanisms that can explain "large distances" (such as getting out of water;, or passage from bacteria to reptiles and then to birds). This is because for those "larger" distances to be traversed, you need so *many simultaneous complex adaptations.* Pierre Grasse[179] said: "explain to me the eye, and I would give you all the rest" (We don't know of any intermediary between the pigmented spot of some invertebrates and the eye, absolutely differentiated, as with the vertebrates and some cephalopods). Drost stated that you need *all* of the factors to be present at the same time; a gradual change would not be possible. If evolution did take place, there would have been many large leaps, qualitative and quantitative, that took place, certainly not gradually.

Mere randomness certainly does not explain evolution at such a fundamental level. If some evolution did take place at that level, then there might be some other laws that we still have not been able to identify. There might be some "programs" that control all of that, and that the Darwinian and all other existent theories cannot explain.

The fossil records. According to Darwin's theory, evolution takes place through small successive changes, the sum of which causes the passage of one species to the next. If this is the reality, we should not have any difficulty finding footprints of the intermediary forms that trace the passage of one organism to the next. Hence, like other evolutionists, Darwin turned to fossils for evidence. He was hoping to find some perfect gradual chains showing the many transitional forms that led from one species to another. But these successions of fossils were nowhere to be found in the world. As Darwin himself stated, "Innumerable transitional forms must have existed, but why do we not find them embedded in countless numbers in the crust of the earth?"

For example, for fish to become amphibians, the most important body changes to occur are: 1) the transformation of fins into feet, and 2) the development of a pelvis to support the amphibian's weight. On the basis of gradual Darwinian evolution, one should expect a wealth of transitional forms showing the development of the appropriate fins, the loss of others, and the slow strengthening of the pelvic bones. Yet there are no such connecting forms anywhere.

Still, Darwin hoped that with time, we will end up discovering all the stages linking the major species, current ones and historical ones that we knew through fossils. There were still many areas on the earth that had not been explored yet, and that would certainly provide the missing proof. But since then, there have been major disenchantments. Paleontologists did discover some new organisms, but, in the great majority of cases, these organisms were either isolated forms or collateral branches and NOT the elusive intermediaries that everyone had been chasing

Then it was suggested that, in order to find these intermediary links, we may have to go far back in time. New hopes were kindled when, in 1909, the American Charles Doolittle Walcott[180] discovered in British Columbia some animal fossils from the Cambrian era (500 million years), and again, in 1947, when an Australian geologist discovered in Australia some fossils dating back to the Precambrian era (700 million years). But, in both cases, the fossilized rocks delivered organisms that were completely unknown, a double deception, as neither of the finds provided the transitional entities that were so eagerly expected.[181] The same happened every time fossils were found.[182] They were mostly fossils of more distinct species, that have appeared then suddenly disappeared, and not of transitional species. The fossils remain missing in the most important places on the evolutionary tree, and plants and animal species seem to have simply "appeared" on that tree.

Furthermore, each time a new fossil is located, it creates more problems than it resolves.[183] With each discovery, a new intermediary sequence had to be imagined in order to accommodate the new finds in the phylogenetic tree. As a result, instead of getting organized and clarified, the evolutionary scheme kept getting tremendously complicated over the years.

A very interesting recent discovery in Georgia further complicated the picture.[184] Some fossilized skulls were discovered of what seems to be *Homo erectus*. They were the most complete among what has been discovered so far, and their ages were put around the same geological time. But what was really significant about the discovery was the incredible diversity of the specimen, which previously classified them into different species. But, being discovered in the same location and from the same geological time, the specimens are considered belonging to a single population of a single species. Thus, what used to be considered several species (*H. erectus*, *H. alibi*, and *H. rudolfensis*) appear to be one. In other words, what used to be considered several intermediary forms turned out to be only one, further widening the gaps and throwing the hypothesis on human evolution into disarray.

Even in the presence of the exceptional cases, contends Denton, such as with the *Archaeopteryx*, a probable transition between birds and reptiles, we cannot be certain that the place we assign to them in the tree is the correct one. This is because a fossil normally preserves only a portion of the perished organism. In the case of vertebrates, it would only include the portion that was partially mineralized during life, such as teeth and bones (or the chitinous exoskeleton of invertebrates), which would comprise only 1% of the biological identity of the organism. The remaining 99% is found in the soft tissues (skin, muscles, nerves, guts,...), and is lost. For instance, if all marsupials had disappeared from the face of the earth, with only skeletons available, no one could have guessed that the gestation process with these organisms is biologically different from placental mammals (embryological development in marsupials starts in the mother's uterus but is completed in an abdominal incubation pouch). The Tasmanian wolf, for example, very closely resembles the common wolf at the skeletal level, with similar skeletal frame, skull, teeth, etc., yet the former is marsupial while the latter is placental.

Still, some evolutionists pretend that there already exists some gradual succession of some skeletons that would allow us to establish some filiation, where each organism in the series has several common characteristics with the preceding as well as the following organisms. The most famous example is that of the equines, where the various forms clearly indicate an evolutionary line. Some eight intermediary forms bridge the *Eohippus* (the first horse to appear on earth, 60 million years ago) to the present day horse. Denton concedes that such lines do exist and indicate some species evolution. But does it allow us to conclude the continuity of all the species, i.e., the descendance of all living organisms out of a common ancestor? That would be far less credible. If eight forms trace the gap between the Eohippus and the present horse, then astronomical numbers are required to pass from the first living cell to the various living organisms. Yet these links are nowhere to be found. Could they have all mysteriously disappeared without leaving a trace?

Homology. Among the proofs presented by the proponents of evolution, *homology* occupies a strong position. Homologous organs are fundamentally similar but are differentiated to serve different purposes. The most cited case is that of the homologous anterior limbs of terrestrial vertebrates: the human hand, the bird's wing, and the bat's wing. All are constructed according to a similar pentadactyl (five-fingered) plan, with similar bones. For Darwin, this strongly suggests the idea of a common ancestor. Another example of homology is the presence of rudimentary organs. These are underdeveloped organs that have no functional role, but which are homologous to organs that were developed normally in some ancestral groups. An example is the *plica semilunaris* in the human eye, which might be considered a residue of the nictitating membrane, found in some mammals.

Denton[185] claims that the homology observed in the human hand, the bat wing, and the porpoise's fin does not suggest a common ancestry. He interestingly observed that the anterior and posterior limbs of the same species show an identical pentadactyl plan, both morphologically and embryologically. Yet no evolutionist dared to pretend that the posterior limbs descended from the anterior ones. There is no doubt, from the evolutionary perspective itself, that the anterior and posterior limbs appeared independently, the former being derived from the fish pectoral fin, the latter from the pelvic fin. But how can such profound similarity between the anterior and posterior limb of the same species be explained from the Darwinian perspective?

How can a gradual random and unrelated accumulation of structural changes lead to such striking similitude?

Furthermore, the observed morphological homology does NOT reflect similar embryological development or genetic similarity. It would have greatly strengthened the validity of the evolutionary interpretation of homology. But the principle clearly cannot be extended in this way; homology can seldom be extended back into embryology, and homologous structures are often specified by non-homologous genes.

For example, the forelimbs in the newt, lizard, and human developed from totally different trunk segments during embryology. Also, the ways different organs are formed in different insect species during metamorphosis are bewilderingly diverse. It thus seems that homologous structures were arrived at via totally different routes.

Further perplexing to the evolutionist, genes coding for homologous structures are not exactly homologous. De Beer[186] stated, "homologous structures need not be controlled by identical genes and the homology of phenotypes does not imply a similarity of genotypes."

Homology cannot be quoted only when it serves the evolutionary cause and ignored when it does not integrate into the evolutionary scheme. A real proof should apply to all situations; otherwise it loses its value.

C. The Islamic Position

Science is clearly uncertain about evolution. Despite some specific observations, these can hardly be extended beyond the observed itself, and Darwin's theory of evolution remains highly probabilistic, even conjectural. And, no matter what argument is presented by evolutionists, one big reality cannot be ignored. This is the fact that evolution is a time-process; it has not and will never be observed directly. The Qur'an states in Surat Al-Kahf:

$$\text{مَا أَشْهَدتُّهُمْ خَلْقَ السَّمَاوَاتِ وَالأَرْضِ وَلا خَلْقَ أَنفُسِهِمْ}$$

I called them not to witness the creation of the heavens and the earth, nor (even) their own creation.[187]

What we have is a reconstructed model, based on some present observations. This would always introduce major uncertainty, as major errors would be introduced in the process filling in the missing pieces. A very large picture is being reconstructed only out of a few available details. The evolutionary model itself will never be observed in action, hence will never be verified.

Then what does Islam say about evolution? Unlike some other religions, Islam itself mentions nothing concerning the origins of the living species and the causes of their variations, except for the origin of the human, i.e., the creation of Adam. Whatever happened was left to the experts to determine, without the religion stating any clear position. Thus, *if* the evolution of any species (other than the human) were to be proven, it would not conflict with anything in Islam. Islam neither says evolution happened, nor does it say evolution did not happen. Nevertheless, Islam is somewhat clear about the creation of Adam. This might be considered the only area of overlap between religion and science regarding evolution.

The Creation of Adam. The creation of Adam has not been described as conclusively in Islam as we commonly think. Some Qur'anic verses and authentic hadeeths give some descriptions, but, being a matter of Ghaib, much remains unknown regarding *how* Adam was created. Obviously, we cannot extrapolate to fill in the gaps, as the scholars have determined that one has to stop at the text in matters of Ghaib. We accept any authentic text in that regard with "no interpretation and no rejection" (بدون تأويل ولا تعطيل). Still, some verses do give some descriptions:

$$إِذْ قَالَ رَبُّكَ لِلْمَلَائِكَةِ إِنِّي خَالِقٌ بَشَرًا مِن طِينٍ$$

Behold, thy Lord said to the angels: "I am about to create man from clay[188]

$$إِنَّا خَلَقْنَاهُم مِّن طِينٍ لَّازِبٍ$$

We have created them out of a sticky and cohesive clay![189]

$$وَلَقَدْ خَلَقْنَا الْإِنسَانَ مِن صَلْصَالٍ مِّنْ حَمَإٍ مَّسْنُونٍ$$

We created man from sounding (dry) clay, from clay (darkened due to being in the water for a long time) mud moulded into shape[190]

$$خَلَقَ الْإِنسَانَ مِن صَلْصَالٍ كَالْفَخَّارِ$$

He created man from sounding clay like unto pottery[191]

Thus, the Qur'anic text indicates that Adam was created from some soil, made into clay, and then fashioned into his shape, all by the Hands of Allah, so as for Iblees (Lucifer) not to feel greater than him:

$$قَالَ يَا إِبْلِيسُ مَا مَنَعَكَ أَن تَسْجُدَ لِمَا خَلَقْتُ بِيَدَيَّ أَسْتَكْبَرْتَ أَمْ كُنتَ مِنَ الْعَالِينَ$$

He (Allah) said: "O Iblees! What prevents thee from prostrating thyself to one whom I have created with my hands? Art thou haughty? Or art thou one of the high (and mighty) ones?"[192]

In Surat Aal-Imran, Allah likens the creation of Eissa to the creation of Adam:

إِنَّ مَثَلَ عِيسَى عِندَ اللَّهِ كَمَثَلِ آدَمَ خَلَقَهُ مِن تُرَابٍ ثُمَّ قَالَ لَهُ كُن فَيَكُونُ

The similitude of Jesus before Allah is as that of Adam; He created him from dust, then said to him: "Be". And he was.[193]

The hadeeths give more details about the next step in the creation of Adam. In one hadeeth, the Prophet ﷺ said:

لما صور الله آدم في الجنة تركه ما شاء الله أن يتركه فجعل إبليس يطيف به ينظر ما هو فلما رآه أجوف عرف أنه خلق خلقا لا يتمالك

When Allah fashioned Adam in Al-Jannah, He left him for as long as He wished, during which Iblees kept passing by him. When he (Iblees) saw him (Adam) hollow inside, he knew that he will not hold by himself (or will not able to resist him)[194]

Then about 40 years (we do not know what kind of years) after Allah created the body of Adam in that hollow shape, He blew in him from His spirit. At that time the command was given to Iblees and the angels to prostrate to him.

فَإِذَا سَوَّيْتُهُ وَنَفَخْتُ فِيهِ مِن رُّوحِي فَقَعُواْ لَهُ سَاجِدِينَ

"When I have fashioned him (in due proportion) and breathed into him of My spirit, fall ye down in obeisance unto him."[195]

From the above descriptions, it seems that Adam was created in his shape and did not evolve out of another being. The following Qur'anic verse gives further information:

يَا بَنِي آدَمَ لاَ يَفْتِنَنَّكُمُ الشَّيْطَانُ كَمَا أَخْرَجَ أَبَوَيْكُم مِّنَ الْجَنَّةِ

O ye Children of Adam! Let not Satan seduce you, in the same manner as He got your parents out of the Garden[196]

The verse indicates that, before he appeared on this earth, Adam was in Al-Jannah and did not have some ancestors on this earth.

Therefore, in the absence of anything conclusive in science, the Qur'anic and Prophetic texts concerning the creation of Adam must be interpreted based on their apparent and direct meaning. One may thus conclude that, according to the Islamic references, Adam was created and appeared on earth in his shape, and he did not evolve out of another organism.

D. Why Was Darwin so Widely Accepted?

The influence of Darwin has been undeniably immense. In terms of science, it has revolutionized biology, anthropology, sociology, and other fields. But more importantly, Darwin had tremendous impact on the religious and philosophical landscape in the West. Before Darwin, the human being occupied the central position in the natural scheme of things, and God was the origin of everything. All of that was to change.

But if the Darwinian theory lacks scientific merits, why was it, and still is, so widely accepted? I believe the answer lies not in the scientific credibility of the theory, but rather in its implications, in light of the socio-political environment.

Even before Darwin, by the 18th century, the intellectual climate in the West had already drastically changed, fueled by the hopes of natural sciences. The new social environment was ready for Darwin.

There was already a struggle with God and the church, resulting in a conscious (or unconscious) effort to diminish the influence of the church and the religion. Darwin's own struggle may represent in part some of the prevalent attitude toward religion and the church. William Howarth (environmental historian) stated that

"Darwin's doubt about Christianity dates before his trip, due to his encounter with slave-owning Christians and the problem of evil in the world."[197]

His theory then provided the somewhat needed "scientific" basis for rejecting some of the basic tenets of the Bible. The very foundation of the church was thus weakened, pushing God further and further away from the realm of science and life itself. The "God people" could not offer a scientific alternative.

Darwin's theory was to be accepted, regardless. If it was not him, it would have been someone else. Strangely, Darwin himself hurried to publish his "Origin" because he thought he was about to be beaten by his fellow naturalist Alfred Russel Wallace, who had come up with much the same idea of evolution through natural selection[198]. So, if it was not Darwin, it would have been someone else, and the success of his theory was inevitable.[199] It became already widely accepted, even before Gregor Mendel's[18] discovery of the laws of heredity, which are essential to explain it. Interestingly, Mendel's work was taking place during the same years Darwin was writing and publishing, but his work was totally ignored until 1900, by which time Darwin's theory was already solidly established. *Thus, the theory was accepted before being fully explained and understood.*

This is not to say that Darwin was blindly accepted. There were previous scientists that postulated the idea of species evolution, such as the famous French naturalist Jean Lamarck and Darwin's own grandfather, Erasmus Darwin. But Charles Darwin's theory was the first genuine attempt to explain *how* evolution occurred. And there was no scientific alternative. For someone who believes in typology, it was much harder, even impossible, to envisage a

---

[18] Gregor Johann Mendel (1822 – 1884) was an Austrian monk who discovered the basic principles of heredity through experiments in his garden. Though farmers had known for centuries that crossbreeding of animals and plants could favor certain desirable traits, Mendel's pea plant experiments established many of the rules of heredity, now referred to as the laws of Mendelian inheritance.

natural process that can explain diversity on earth. Darwin offered the only explanation.

More significantly, Darwin's theory was the first effort to bring the study of life on Earth fully into the conceptual sphere of science.[200] By the mid-nineteenth century, it became acceptable to everyone, including the rigid religious dogmatists, that everyday physical phenomena were readily explicable in terms of natural causes. Divine intervention in everyday life was becoming less and less credible. Darwin's revolution extended the scientific method to all biological sciences by giving a natural explanation of the design of living things.

As the years passed, the obvious discontinuities in nature could no longer be perceived. Consequently, debate slackened, and there was less need to justify the idea of evolution. Darwin's concepts started permeating every aspect of biological thought so that today all biological phenomena are interpreted in Darwin's terms. Professional biologists are subjected throughout their working life to continued affirmation of the truth of the Darwinian theory. Research, discussions, debates, etc., assumed the truth of the Darwinian theory and kept continuously reinforcing its credibility. Whole "scientific" fields were based on it (e.g., phylogeny). It became almost impossible for someone who is already seeing everything in light of that theory to question it or see things otherwise. I further believe that the more specialized training of the modern scientists limits their ability to reexamine the bigger picture. If it is very difficult for someone who was raised in a certain religion to change, it is almost impossible for someone who has been *scientifically* seeing the world through the Darwinian theory to observe it otherwise. One would have to challenge the whole scientific world, which appears to be no less dogmatic than the religious world.

Eventually, Darwin's theory was elevated from a highly speculative hypothesis to a dogma. Neo-Darwinists became more Darwinistic than Darwin himself. Julian Huxley[19], an English

---

[19] Julian Huxley (1887 – 1975), English biologist, philosopher, educator,

evolutionary biologist, stated at a conference in 1959: "the first point to make about Darwin's theory is that it is no longer a *theory*, but a *fact*... we are no longer having to bother about establishing the *fact* of evolution..." The famous evolutionary biologist Richard Dawkins[20] even more strongly stated that: "The theory is about as much in doubt as the earth goes around the sun." The big gaps in nature that Darwin himself acknowledged became completely invisible to his followers, and any opponent of Darwin is automatically dismissed as religiously motivated, and lacking in scientific credibility.

The impact of the theory was not restricted to science but extended to the whole philosophy and ethics of the West. These are now founded largely on the premise that the human was born out of a purely blind and random molecular selection. Thus, the whole of the 20th century would be incomprehensible without the Darwinian revolution. Some social and political currents that swept through the whole world, such as Marxism and all its practices, would have been impossible without Darwinism's intellectual affirmation.[201]

But, in the midst of all of the above, Darwin himself had some serious doubts. He admitted himself that there was no evidence that any of the major divisions in nature had been crossed in a gradual manner.[202] By the time of the publication of the last edition of the "Origin" in 1872, he became plagued with self-doubt and frustrated with his inability to meet the many challenges and objections raised against his theory by his opponents. But, by then, his book had already widely spread and the theory was largely accepted, more strongly by others than by him. That is why rather than an atheist, which describes many Neo-Darwinists,

---

and author who greatly influenced the modern development of embryology, systematics, and studies of behavior and evolution.

[20] Clinton Richard Dawkins (born 1941) is an English ethologist, evolutionary biologist, and writer. He is an emeritus fellow of New College, University of Oxford.

Darwin himself could be categorized as more agnostic, still eagerly searching for the truth.

5. Science as a Tool to Establish Belief

The relationship between science and religion is commonly perceived to be that of a conflict, though the reality is not such. Blanket statements are commonly made by people who have either not accurately examined science, religion or both. Generalizations are commonly made by people with partial exposures to either or both.

But, if science is certain and religion is true, the two would intersect.

In Islam, emphasis is placed on the continuity of life, where all aspects are parts of the same whole.

قُلْ إِنَّ صَلَاتِي وَنُسُكِي وَمَحْيَايَ وَمَمَاتِي لِلَّهِ رَبِّ الْعَالَمِينَ

Say: "Truly, my prayer and my service of sacrifice, my life and my death, are (all) for Allah, the Cherisher of the Worlds.[203]

Life, all of life, has one main purpose that all parts are supposed to serve.

وَمَا خَلَقْتُ الْجِنَّ وَالإِنسَ إِلاَّ لِيَعْبُدُونِ

I (God) have only created Jins and men, that they may worship Me.[204]

As such, everything becomes a tool to ultimately achieve the main purpose, to worship God, and science itself becomes a tool of worship.

The famous Muslim philosopher Averroes (إبن رُشْد) indicated that philosophy and science seek to know the truth, and, if accurate, they would lead to the same result as the religion. As such, science and research become means to know God.

إِنَّ فِي خَلْقِ السَّمَاوَاتِ وَالْأَرْضِ وَاخْتِلَافِ اللَّيْلِ وَالنَّهَارِ لَآيَاتٍ لِّأُولِي الْأَلْبَابِ

الَّذِينَ يَذْكُرُونَ اللَّهَ قِيَامًا وَقُعُودًا وَعَلَىٰ جُنُوبِهِمْ وَيَتَفَكَّرُونَ فِي خَلْقِ السَّمَاوَاتِ وَالْأَرْضِ رَبَّنَا مَا خَلَقْتَ هَٰذَا بَاطِلاً سُبْحَانَكَ فَقِنَا عَذَابَ النَّارِ

Behold! In the creation of the heavens and the earth, and the alternation of night and day,- there are indeed Signs for men of understanding;

Men who celebrate the praises of Allah, standing, sitting, and lying down on their sides, and contemplate the (wonders of) creation in the heavens and the earth, (With the thought): "Our Lord! not for naught Hast Thou created (all) this! Glory to Thee! Give us salvation from the penalty of the Fire."[205]

The connection between science or any form of knowledge and religion was established in the very first revealed Qur'anic verse. The command links knowledge to faith at the most basic level:

إقرأ بِسمِ رَبِّك

Read in the name of your Lord![206]

It is a call to seek knowledge, *in the name of God.* A knowledge severed from faith leads to arrogance, aggression, even destruction.

As such, the relationship between science and religion is that of mutuality. *Correct knowledge and good science lead to God and religion, while science needs the guidance of religion.* It did happen before, historically, with the Muslim scientists, and it is badly needed today.

CONCLUSION

Belief is a necessity for people's life. It gives significance to their life, and life to their heart. It helps them achieve levels of happiness that would not be possible otherwise. It gives them the feeling of fulfillment in knowing that they are doing what they are meant to do. Without belief, one would be more deprived and less fortunate than animals (not meant to be insulting), for as Al-Jisr[207] beautifully stated:

- Animals starve like we do, but are not plagued with worries concerning future sources of sustenance.

- They beget children like we do, but are guarded against the sadness of losing a parent or a child.

- They get sick like we do, but they are guarded against the fear of what sickness might lead to, the agony of death, and leaving loved ones behind.

- They kill to eat or to defend themselves, but never kill just out of aggression or hate.

The human is subject to all of the above worries and evil tendencies. And these can only be tamed by a strong belief. Specifically, the Islamic belief is comprehensive enough to meet all the above challenges and more. It is a knowledge, an intention, a belief, and an action, balanced and deep enough to have such an impact.

The principal means to reach that belief is the person's own mind and thinking. *Man is a servant to his own thinking before he is a servant to his Lord, and can only become a servant to His Lord through his thinking.*[208] It is what God favored the human with, for it is the very tool he can know Him with.

Knowledge and the various worldly sciences become tools used by the mind to learn about God and to strengthen one's belief. If they are accurate and the religion is true, they should lead to the same conclusions.

But key factors in approaching such knowledge are humbleness and sincerity. The very first revealed verse commands the believer to acquire knowledge, but to do it in connection with the Lord:

$$ اقْرَأْ بِاسْمِ رَبِّكَ $$

Read in the name of thy Lord[209]

A knowledge severed from the Lord and pursued for the wrong reason would lead to arrogance, and it blinds the person to his mind's limitations. Then the very tool that was meant to guide the person to the truth and ensure happiness becomes a tool of misguidance, confusion, and misery. A sharp mind with partial knowledge and wrong motives can lead to confusion.

The Qur'an gives us examples of how an arrogant mind can be blinded even to the most basic realities, as in these verses:

$$ أَوَلَمْ يَرَ الإِنسَانُ أَنَّا خَلَقْنَاهُ مِن نُّطْفَةٍ فَإِذَا هُوَ خَصِيمٌ مُّبِينٌ $$

$$ وَضَرَبَ لَنَا مَثَلاً وَنَسِيَ خَلْقَهُ قَالَ مَنْ يُحْيِي الْعِظَامَ وَهِيَ رَمِيمٌ $$

$$ قُلْ يُحْيِيهَا الَّذِي أَنشَأَهَا أَوَّلَ مَرَّةٍ وَهُوَ بِكُلِّ خَلْقٍ عَلِيمٌ $$

Doth not man see that it is We Who created him from a droplet? Yet behold! He suddenly (stands forth) as an open adversary!

And he makes comparisons for Us, and forgets his own (origin and) Creation: He says, "Who can give life to (dry) bones and decomposed ones (at that)?"

Say, "He will give them life Who created them for the first time! for He is Well-versed in every kind of creation![210]

A futile argument is given by someone who forgot his own start as a droplet, becoming suddenly argumentative, and who overlooked his own creation the first time, arguing about the impossibility of a second creation!

The mind is limited also, and often by the lack of sufficient knowledge. No mind has such a complete and comprehensive knowledge of everything in the universe as to claim to have the absolute truth on its own. Our knowledge is always partial and might easily be missing some essential elements.

The mind's limitation is particularly major when delving into the metaphysical world. We are so bound and restricted by our own world that we commonly, and often unconsciously, assume that everything existent is bound by physical and temporal dimensions. Thus, when considering the Unseen on our own, we either miscomprehend it or simply dismiss it as unreal. We assume we know of all the existent "dimensions," and we find it implausible, or even impossible, for something to exist beyond this world.

But, despite being beyond the direct grasp of the mind, the metaphysical world as described in Islam cannot be irrational. It includes that which the mind cannot attain on its own, but *not* that which is contradictory to the sound mind, i.e., impossible or illogical. To indicate, for example, that the part is larger than the whole, or that a multiple equals one of the same category would be rejected, and no one can demand for it to be accepted under the pretext that it is metaphysical. The truth cannot contradict sound logic. Ibn Taymiyah[211] puts it eloquently in his statement:

نحن نعلم أن الرسل لا يخبرون بمحالات العقول بل بمحارات العقول، فلا يخبرون بما يعلم العقل انتفاءه، بل يخبرون بما يعجز العقل عن معرفته.

We know that the Messengers do not bring forth that which is irrational, but rather, that which is supra-rational; thus, they do not deliver that which the mind can reject as impossible, but rather that which the mind cannot attain on its own.

For all of the above reasons, as solid as one's belief might be, it is never constant. Confusions, misinformation, misconceptions, and biases might all lead to doubts and weaken one's belief. Real doubts must be addressed, as significant doubts could eventually lead to disbelief. But the occurrence of certain levels of doubt through life should be expected; it is the nature of a realistic belief. There will always be some lack of certainty, which does not mean lack of belief.

Moments of strong belief (certitude?) commonly alternate with doubts. The focus needs to be placed on these moments of belief, instead of the moments of doubt. Though not permanent, these periods of strong belief tell the person that his belief is present, that he is a believer. He should accept that, then move on, and not remain stuck in the in-between place, where he is not sure what to consider himself, a believer or a non-believer.

Subsequently, whatever actions he takes will reinforce these moments of belief, and belief would become his reality.

This is because Eeman/belief is not an exclusive experience of the mind. It involves feelings, habits, rituals,… Giving the mind too much emphasis in the whole "equation of belief" can get in the way of establishing a strong belief. A doubt in the intellectual component does not indicate a doubt in the belief itself.

As such, the initial doubt itself can be a motivation for further research and learning, to eventually reinforce the belief and to keep it alive.[1] As Frederick Buechner[2] stated, "Doubts are the ants in the pants of faith. They keep it awake and moving."

---

[1] It would be very beneficial to examine the relationship between doubt or atheism and the various personality traits and disorders. An in-depth analysis of such correlation would help identify the real causes of the issues one is dealing with, which is crucial in addressing them.

[2] Carl Frederick Buechner (born 1926) is an American writer and Presbyterian theologian.

Muhammad Al-Ghazali[3] remarked that historically, "Ilm Al-Kalam" (the traditional academic way of studying Islamic creed) has reduced the study of Islamic Belief into a purely academic exercise, where conclusions are drawn the way calculators give results.[212] It trains the mind in a purely intellectual way, keeping it severed from the heart. The same approach is reinforced yet further in today's society, where the weight is placed on intellect, information, and eloquence.

While intellectual knowledge is certainly crucial, it is only the beginning of the path toward belief. One may read the best book on proving the Existence of God, but, unless that knowledge is established in his heart, it will not develop into a belief. He would still default to what he already had in his heart, every time he is faced with a challenge. A belief is not *only* built intellectually, though it may start as such.

Instilling such a solid belief in the heart is best done by following the Qur'anic way. It connects every observed event and experience with God, such as in the verses below.

- When experiencing a blessing:

$$وَمَا بِكُم مِّن نِّعْمَةٍ فَمِنَ اللَّهِ$$

- *And any good thing you have is but from Allah*[213]
- When exposed to something fascinating:

$$وَلَقَدْ خَلَقْنَا الْإِنسَانَ مِن سُلَالَةٍ مِّن طِينٍ$$

$$ثُمَّ جَعَلْنَاهُ نُطْفَةً فِي قَرَارٍ مَّكِينٍ$$

$$ثُمَّ خَلَقْنَا النُّطْفَةَ عَلَقَةً فَخَلَقْنَا الْعَلَقَةَ مُضْغَةً فَخَلَقْنَا الْمُضْغَةَ عِظَامًا$$

$$فَكَسَوْنَا الْعِظَامَ لَحْمًا ثُمَّ أَنشَأْنَاهُ خَلْقًا آخَرَ فَتَبَارَكَ اللَّهُ أَحْسَنُ الْخَالِقِينَ$$

---

[3] Mohammed al-Ghazali al-Saqqa (1996—1917);(الشيخ محمد الغزالي السقا) was an Egyptian Islamic cleric and scholar, widely credited with contributing to the Islamic revival in Egypt over the last decade.

Man We did create from a quintessence (of clay);
Then We placed him as (a drop of) sperm in a place of rest, firmly fixed;
Then We made the sperm into a clot of congealed blood; then of that clot We made a (fetus) lump; then we made out of that lump bones and clothed the bones with flesh; then we developed out of it another creature. **So blessed be Allah, the best to create!**[214]

- When going through a hardship:

$$\text{أَمَّن يُجِيبُ الْمُضْطَرَّ إِذَا دَعَاهُ وَيَكْشِفُ السُّوءَ}$$

- *Or, Who listens to the (soul) distressed when it calls on Him, and Who relieves its suffering?*[215]

- When feeling distressed:

$$\text{الَّذِينَ آمَنُواْ وَتَطْمَئِنُّ قُلُوبُهُم بِذِكْرِ اللَّهِ أَلاَ بِذِكْرِ اللَّهِ تَطْمَئِنُّ الْقُلُوبُ}$$

- *Those who believe, and whose hearts find satisfaction in the remembrance of Allah. for without doubt in the remembrance of Allah do hearts find satisfaction.*[216]

- When experiencing temptations:

$$\text{وَإِلاَّ تَصْرِفْ عَنِّي كَيْدَهُنَّ أَصْبُ إِلَيْهِنَّ وَأَكُن مِّنَ الْجَاهِلِينَ}$$

- *(Yussuf said) "Unless Thou turn away their snare from me, I should (in my youthful folly) feel inclined towards them and join the ranks of the ignorant."*[217]

- When giving charity:

$$\text{إِنَّمَا نُطْعِمُكُمْ لِوَجْهِ اللَّهِ لا نُرِيدُ مِنكُمْ جَزَاء وَلا شُكُورًا}$$

- *(Saying), "We feed you for the sake of Allah alone: no reward do we desire from you, nor thanks."*[218]

- When standing up for justice:

$$\text{يَا أَيُّهَا الَّذِينَ آمَنُواْ كُونُواْ قَوَّامِينَ لِلّهِ شُهَدَاء بِالْقِسْطِ وَلاَ يَجْرِمَنَّكُمْ شَنَآنُ قَوْمٍ عَلَى أَلاَّ تَعْدِلُواْ اعْدِلُواْ هُوَ أَقْرَبُ لِلتَّقْوَى وَاتَّقُواْ اللّهَ إِنَّ اللّهَ خَبِيرٌ بِمَا تَعْمَلُونَ}$$

130

- O ye who believe! stand out firmly for Allah, as witnesses to fair dealing, and let not the hatred of others to you make you swerve to wrong and depart from justice. Be just: that is next to piety: and fear Allah. For Allah is well-acquainted with all that ye do.[219]

The crucial formula is to always connect different aspects of our lives with the Guidance and Supervision of Allah, recognizing all blessings and support as coming from Him. This was expressed in the declaration of Prophet Ibrahim in the Qur'an when asked about his Lord:

الَّذِي خَلَقَنِي فَهُوَ يَهْدِينِ

وَالَّذِي هُوَ يُطْعِمُنِي وَيَسْقِينِ

وَإِذَا مَرِضْتُ فَهُوَ يَشْفِينِ

وَالَّذِي يُمِيتُنِي ثُمَّ يُحْيِينِ

وَالَّذِي أَطْمَعُ أَن يَغْفِرَ لِي خَطِيئَتِي يَوْمَ الدِّينِ

The One Who created me, and it is He Who guides me;
Who gives me food and drink,
And when I am ill, it is He Who cures me;
Who will cause me to die, and then to life (again);
And who, I hope, will forgive me my faults on the Day of Judgment.[220]

Whatever we experience would then reinforce our belief, burying our doubts deeper and deeper under a *living* belief.

But, while the above is the ideal belief one is aiming for and continuously striving to attain, anything short of that still counts. At times, especially when faith or conviction decreases, other factors might play a bigger role, such as a good company, as the Prophet ﷺ recommended in the hadeeth:

عَلَيْكُمْ بِالْجَمَاعَةِ وَإِيَّاكُمْ وَالْفُرْقَةَ فَإِنَّ الشَّيْطَانَ مَعَ الْوَاحِدِ وَهُوَ مِنَ الِاثْنَيْنِ أَبْعَدَ

Adhere to the group and avoid divisions, for the devil is with the singular one, and is further from the two.[221]

Or hope in some worldly gain, when the motivation to please Allah or Al-Jannah may decrease:

وَمَا أَنْفَقْتُمْ مِنْ شَيْءٍ فَهُوَ يُخْلِفُهُ وَهُوَ خَيْرُ الرَّازِقِينَ

And nothing do ye spend in the least (in His cause) but He replaces it: for He is The Best of those who grant sustenance.[222]

إِنَّ الْمُصَّدِّقِينَ وَالْمُصَّدِّقَاتِ وَأَقْرَضُوا اللَّهَ قَرْضًا حَسَنًا يُضَاعَفُ لَهُمْ وَلَهُمْ أَجْرٌ كَرِيمٌ

For those who give in Charity, men and women, and loan to Allah a Beautiful Loan, it shall be increased manifold (to their credit), and they shall have (besides) a liberal reward.[223]

As such, Eeman/belief becomes *dynamic*, with different factors having different significance in different times, continuously eying the ideals, but firmly grounded in reality.

And reality dictates that challenges will always happen in life. It must be expected, as indicated in the Qur'an.

أَحَسِبَ النَّاسُ أَن يُتْرَكُوا أَن يَقُولُوا آمَنَّا وَهُمْ لا يُفْتَنُونَ

Do men think that they will be left alone on saying, "We believe", and that they will not be tested?[224]

But, when challenges and tests do happen, they should not completely derail one's belief. A balanced and dynamic belief should accommodate these tests, which become tools of growth themselves.

But challenges are not limited to hardships. One such serious challenge is the ability to resist some prevalent ways of thinking, strongly promoted in today's "modern" world. We are continuously driven to ask all sorts of questions, important and unimportant ones, without taking into account our limited abilities. We refuse to accept that some questions simply do not have available answers.

But why do we have to have all the answers, anyway? Maybe "not knowing" is what makes us human. Not being expected to know everything brings relief!

We still aim for certitude yet realistically accept a "most likelihood." We seek to know and to understand, but only that which can be understood. And until we fully understand, we do not have to put our belief on hold. We have enough to lean against to move forward.

During one of my trips to Lebanon, a dear brother wrote to me complaining about his shaky belief after some difficulties he experienced. I wrote him back:

Yesterday, I decided to take a walk around the Palestinian refugee Camp of Sabra in Beirut, an incredibly poor neighborhood. I expected to see poverty, yet it was still shocking to say the least, sharply contrasting with the rest of the metropolitan city, only a few streets away. I thought about my fragility in the face of life's challenges, even with my much easier circumstances.

Despite the incredibly difficult life circumstances (by all standards), people in the camp still have an amazing resiliency and a solid feeling of contentment. You clearly see it in people's faces, and the camp's bustling life.

I wish I had that myself.

*I think part of the problem is that we live in an environment where everything is guaranteed. Hence, we take much for granted, feeling entitled to what we desire. Everything to us becomes a strict relationship of cause/effect. If we do the "cause," then we **demand** the "effects." We do the cause with the rigid expectation of the effect. When we sacrifice something for the sake of God, we **demand** an immediate and specific reward, for compensation.*

And when we do not see the reward we expected (i.e., when real life happens), we feel disappointed, and we start questioning God. "Why?" We demand to know, but soon we find out that we can't know. Then we question our belief itself!

Life is full of these challenging moments, and knowing what to expect helps us deal with those moments, while learning at the same time.

"Life" becomes an experience.

Being realistic in what one expects of himself to know is fundamental. We should still aim high and pursue any knowledge we can attain, but at the same time be realistic in our expectations and accept the fact that the ideal is just that, an ideal, and should not be the minimal accepted standard.

We pray for God the Almighty to enlighten our heart and our life with the truth.

We repeat the supplication of the great Prophet ﷺ, praying for Allah to **give us enough certitude to make easy the problems and challenges in life!**

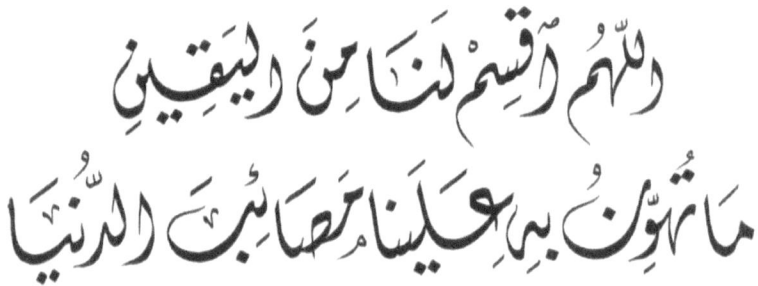

APPENDIX:

BOOK DISCUSSION QUESTIONS

1. How does the author differentiate between "faith" and "certainty"? Can belief exist without full certainty?
2. What role does the concept of *fitrah* (natural disposition) play in the formation of belief, according to the book?
3. How are doubt and satanic whispers distinguished in the text, and why is this distinction important?
4. In what ways does the author argue that science can be used to support, rather than contradict, belief in God?
5. What are the various paths outlined by the author through which a person can arrive at belief?
6. How does the author address the apparent conflict between evolution and Islamic belief?
7. According to the book, how does the modern, materialistic world affect the human capacity to believe?
8. How can belief be protected and strengthened in environments that promote skepticism or relativism?
9. What does the author suggest is the most effective way to address and overcome doubts in one's faith?
10. How does the book challenge common misconceptions about the relationship between Islam and rational inquiry?
11. If you could ask the author one question about this book, what would it be?
12. On a scale of 1 to 5, with 5 being the highest, how would you rate this book? Why did you give it that rating?

135

REFERENCES

Sources in English and French

Adler, J. 2005. Evolution of a Scientist. Newsweek. Dec. 12.

Ahmad, S.F. & S.S. Ahmad. 2004. *God, Islam, and the Skeptic Mind*. 238 pp. Blue Nile Publishing.

Ali, Abdullah Yusuf. 1984. *The Holy Qur-an: English Translation of the Meanings and Commentary*. King Fahd Holy Qur-an Printing Complex, Madinah, Saudi Arabia.

al-Mehri, A.B. "Doubt and Certainty." *The Quran Project*. Accessed May 21, 2025. https://www.quranproject.org/Doubt-and-Certainty-661-d.

Asad, M. 2003. *The Message of the Qur'an*. 1200 pp. The Book Foundation.

Baird, J. 2014. "Doubt as a Sign of Faith." New York Times.

Bowlby, J. 1992. *Charles Darwin: A New Life*. 528 pp. W.W. Norton & Company.

Bucaille, M. 2003. *The Bible, the Qur'an and Science: The Holy Scriptures Examined in the light of Modern*

Knowledge. 272 pp. Tahrike Tarsile Qur'an.

Cambridge University Press. "Doubt." *Cambridge Dictionary.* Accessed May 20, 2025. https://dictionary.cambridge.org/us/dictionary/english/doubt.

Crystal, D. 2003. *The Cambridge Encyclopedia of the English Language.* 506 pp. Cambridge University Press.

Darwin, C. 2003. *The Origin of Species* 150th Ed. 576 pp. Signet Classics.

Dawkins, R. 1976. *The Selfish Gene.* Oxford University Press.

De Beer, G. 1971. *Homology, an unresolved problem.* Oxford University Press.

Denton, M. 1986. *Evolution: A Theory in Crisis.* 368 pp. Adler & Adler Publishers.

Denton, M.J. 1998. *Nature's Destiny: How the Laws of Biology Reveal Purpose in the Universe.* 454 pp. The Free Press.

Dictionary.com. *Dictionary.com*, Random House, accessed May 19, 2025. https://www.dictionary.com/browse.

Dostoyevsky, F. 2002 (Russian original 1879). *The Brothers Karamazov.* 824 pp. Publisher: Farrar, Straus and Giroux.

Gegax, T.T., J. Raymond, J. Sieder, J. Reno, K. Skipp. 2005. "Doubting Darwin: How Did Life, in its Infinite Complexity, Come to Be?" Newsweek. Feb. 7.

Grousset, V. 1991. "Enigme: Les vertiges de la science." Le Figaro Magazine. 23 Fevrier. pp. 74—77.

Hart, M. 2000. *The 100: A Ranking of the Most Influential Persons in History.* 556 pp. Citadel Publishers

Hitching, F. 1982. "Was Darwin wrong?" Life. April 1982. V. 5, No. 4: 48-52

Jones, M. 2008. "Who Was More Important: Lincoln or Darwin?" Newsweek. July 7.

Kolodiejchuk, B. (Ed.) 2007. *Mother Teresa: Come Be My Light: The Private Writings of the Saint of Calcutta.* 416 pp. Doubleday Pub.

Krakauer, J. 2007. "Into the Wild." 224 pp. Anchor Books.

Lyell, C. 1998. *Principles of Geology: Being an Attempt to Explain the Former Changes of the Earth's Surface by Reference to Causes now in Operation.* p. 528. Penguin Classics.

Margenau, H. & R. A. Varghese (Eds.). 1993. *Cosmos, Bios, Theos: Scientists Reflect on Science, God, and the Origins of the Universe, Life, and Homo sapiens.* p. 285 Open Court Publishing Company.

Merriam-Webster. Merriam-Webster.com Dictionary. Accessed May 21, 2025. http://www.merriam-webster.com/dictionary

Ontario Consultants on Religious Tolerance. "Glossary of Religious Terms: P–R." *ReligiousTolerance.org.* Accessed May 21, 2025. https://www.religioustolerance.org/gl_r.htm

Ortoli, S. 1987. "L'évolution contestée." Science et Vie. No. 834: pp. 42-54, 162.

Pecorino, Philip A. "Requirements of a Definition." *Philosophy of Religion: Textbook by Dr. Philip A. Pecorino.* Queensborough Community College, CUNY. Accessed May 21, 2025. https://www.qcc.cuny.edusocialSciences/ppecorino/PHIL_of_RELIGION_TEXTCHAPTER_10_DEFINITION/Requirements_of_a_Definition.htm.

Pew Research Center. "The Global Religious Landscape: Executive Summary." *Pew Forum on Religion & Public Life.* December 18, 2012. Accessed May 22, 2025. https://www.pewforum.org/2012/12/18/global-religious-landscape-exec/.

Pilorge, T. 1993. "*Pinsons de Darwin: Non a la xenophobie.*" Science et Vie. # 906. 84-87.

Pope, Hugh. "Faith." *The Catholic Encyclopedia.* Vol. 5. New York: Robert Appleton Company, 1909. Accessed May 21, 2025. https://www.newadvent.org/cathen/05752c.htm.

"Religion." *Academic Kids Encyclopedia.* Accessed May 21, 2025. https://academickids.com/encyclopedia/index.php/Religion

"Religion." *ScienceDaily.* Accessed May 21, 2025. http://www.sciencedaily.com/articles/r/religion.htm.

Sabra, A. I. 2003. "Ibn al-Haytham: Brief life of an Arab mathematician." Harvard Magazine, September-October, 2003.

Samford University. *Center for Science and Religion*. Accessed May 21, 2025. http:/www.samford.edu/scienceandreligion/.

Sample, I. 2013. "Skull of *Homo erectus* Throws Story of Human Evolution in Disarray." The Guardian, October 17.

Sarton, G. 1975. *Introduction to the History of Science*: V. 1: From Homer to Omar Khayyam." 840 pp. Krieger Pub Co.

Schatzman, M. 1990. *The Origin of Darwin's Despair.* New Scientist. 1729.

Staff. 2005. "The Mystery of the Origin of the Species." Newsweek. Dec. 12.

Staune, J. 1991. "L'Evolution condamne Darwin." Le Figaro Magazine. No. 587. pp. 74—84.

Taunton, Larry Alex. "Listening to Young Atheists: Lessons for a Stronger Christianity." *The Atlantic*, June 6, 2013. Accessed May 22, 2025. https://www.theatlantic.com/national/archive/ 2013/06/listening-to-young-atheists-lessons-for -a-stronger-christianity/276584/.Patheos+4

The Free Dictionary. "Doubt." *The Free Dictionary by Farlex* accessed May 18, 2025. https://www.thefreedictionary.com/ doubt.

Wells, J. 2002. *Icons of Evolution.* p. 338. Regnery Publishing.

Sources in Arabic

.القرآن الكريم

(Title in English: The Holy Qur'an)

ابن تيمية، أحمد بن عبد الحليم بن عبد السلام. 2009. "درء تعارض
العقل والنقل." دار الكتب العلمية، بيروت، لبنان.

(Title in English: Averting the Conflict between Reason and Revelation)

ابن العثيمين، محمد بن صالح. 2014. "مجموع فتاوى ورسائل العثيمين."
دار الثريا للنشر، الرياض، المملكة العربية السعودية.

(Title in English: The Collection of Fatwas and the Messages of Ibn Uthaimeen)

ابن قيّم الجوزيّة، محمد بن أبي بكر بن أيوب. 1992. "الطبّ النّبوي."
دار مكتبة الهلال، بيروت، لبنان.

(Title in English: The Prophetic Medicine)

ابن قيّم الجوزيّة، محمد بن أبي بكر بن أيوب. 1997. "تهذيب مدارج
السالكين." دار مؤسسة الرسالة، بيروت، لبنان.

(Title in English: The Steps of the Followers)

ابن كثير، أبو الفداء عماد الدين إسماعيل. 2002. "تفسير القرآن
العظيم." الجزء السابع. دار طيبة، الرياض، المملكة العربية السعودية.

(Title in English: The Interpretation of the Great Qur'an)

ابن كثير، أبو الفداء عماد الدين إسماعيل. 2010. "البداية والنهاية."
الجزء السابع. دار ابن كثير، بيروت، لبنان.

(Title in English: The Beginning and the End)

البخاري، محمد بن اسماعيل. " الجامع الصحيح." دار البشائر الاسلامية، بيروت، لبنان.

(Title in English: The Authentic Compilation)

بعيون، سهى. 2008. "إسهام العلماء المسلمين في العلوم في الأندلس." دار المعرفة، بيروت، لبنان.

(Title in English: The Contribution of the Muslim Scholars to Sciences in Andalusia)

بعيون، سهى. 2014. " إسهام المرأة الأندلسية في النشاط العلمي في الأندلس." الدار العربية للعلوم ناشرون، بيروت، لبنان.

(Title in English: The Contribution of the Andalusian Woman to the Scientific Activities in Andalusia)

البنّا، حسن. 1990. " مجموعة رسائل الإمام الشهيد حسن البنّا." دار الدعوة، الإسكندرية، مصر.

(Title in English: The Collection of the Messages of Imam Shaheed Hassan Al-Banna)

الجسر، نديم. 1969. " قصة الايمان بين الفلسفة والعلم والقران." دار العربية، طرابلس، لبنان.

(Title in English: The Story of Belief, in Lights of Philosophy, Science, and the Qur'an)

الحميري، عبد الملك ابن هشام. 1990. "السيرة النبوية لإبن هشام." دار الجيل، بيروت، لبنان.

(Title in English: The Prophetic Biography of Ibn Hisham)

زيدان، عبد الكريم. 2004. " الوَجيز في أصول الفقه." مؤسسة الرسالة، بيروت، لبنان.

(Title in English: Abbreviation of the Principles of Islamic Jurisprudence)

السكاف، اسعد نصر الله ومحمود مطرجي. 1988. " تاريخ العلوم عند العرب." دار نظير عبود، بيروت، لبنان.

(Title in English: History of Sciences of the Arabs)

الشافعي، محمد بن ادريس. 1990. " كتاب الأم." دار المعارف، بيروت، لبنان.

(Title in English: The Mother Book)

العودة، سلمان. 2013. " عقلي المؤمن".

www.islamtoday.net/salman/services/printart-28-127197.html
(Title in English: My Believing Mind)

الغزالي، ابو حامد محمد. 2004. " إحياء علوم الدين." الجزء الخامس، دار المعرفة، بيروت، لبنان.

(Title in English: The Revival of the Religious Sciences)

الغزالي، محمد. 1980. " عقيدة المسلم." دار التراث العربي، القاهرة، مصر.

(Title in English: The Creed of the Muslim)

القرضاوي، يوسف. 1991. "الإيمان والحياة. " مؤسسة الرسالة، بيروت، لبنان.

(Title in English: Belief and Life)

معجم المعاني الجامع:

http://www.almaany.com /ar/dict/ar-ar/ولي/

معجم المعاني الجامع:

http://www.almaany.com /ar/dict/ar-ar/الفطرة/

الموسوعة العقدية:

http://www.dorar.net/enc/aqadia/3162

الميداني، عبد الرحمن حسن حبنّكة. 1992. "صراع مع المَلاحِدة حتى العَظْم." دار القلم، دمشق، سوريا.

(Title in English: A Debate with the Atheists to the Bones)

الميداني، عبد الرحمن حسن حبنّكة. 1994. "العقيدة الاسلامية واسسها." دار القلم، دمشق، سوريا.

(Title in English: The Islamic Creed and its Foundations)

هونكة، زيغريد. 1993. " شمس العرب تسطع على الغرب." دار الجليل، بيروت، لبنان.

(Title in English: The Sun of the Arabs Shines over the West)

ABOUT THE AUTHOR

Dr. Imad Bayoun is widely engaged in religious education, particularly among youth, as a lecturer for the Muslim American Society (MAS).

Born in Beirut, Lebanon, Dr. Bayoun earned his B.S. and M.S. from the American University of Beirut. Then he moved to the US and completed a Ph.D. in Entomology at Texas A&M University. He currently resides in Riverside, California, where he works as the Insectary and Quarantine Officer in the Department of Entomology at the University of California in Riverside.

Dr. Bayoun also holds a Ph.D. in Islamic Studies from the Graduate Theological Foundation. He has taught courses such as Aqeedah and Fiqh as-Seerah (Jurisprudence of the Prophet's Biography) at the Islamic American University. Some of his recorded lectures on topics including Purification of the Heart, Muslim Character, and others have benefited a wide audience and are available for streaming or download at imadbayoun.com.

NOTES

[1] Krakauer, J. 2007. *Into the Wild*. 224 pp. Anchor Books.

[2] القرضاوي، يوسف. 1991 ."الإيمان والحياة." مؤسسة الرسالة، بيروت، لبنان
(Title in English: Belief and Life).

[3] Dostoyevsky, F. 2002 (Russian original 1879). *The Brothers Karamazov*. p. 824.

[4] Pew Research Center, "The Global Religious Landscape: Executive Summary," Pew Forum on Religion & Public Life, December 18, 2012, accessed May 22, 2025.

[5] Ibid.

[6] Surat Ibrahim, verse 10.

[7] Surat An-Nissaa, verse 134.

[8] Surat An-Nahl, verse 30.

[9] ابن كثير. أبو الفداء عماد الدين إسماعيل. 2010. "البداية والنهاية." الجزء السابع. دار ابن كثير، بيروت، لبنان
(Title in English: The Beginning and the End).

[10] Hadeeth narrated by Anas Bin Malik in *Sahih Bukhari*, no. 4091.

[11] Surat Aal-Imran, verse 185.

[12] "وَلِيّ," معجم المعاني الجامع, Almaany.com, accessed May 22, 2025.

[13] Surat Al-Insaan, verse 3.

[14] الميداني. عبد الرحمن حسن حبنكة. 1994. "العقيدة الاسلامية واسسها." دار القلم. دمشق، سوريا
(Title in English: The Islamic Creed and its Foundations).

[15] معجم المعاني الجامع almaany.com, accessed May 22, 2025.

[16] Surat Al-Zukhruf, verse 27.

[17] Surat Al-Room, verse 30.

[18] Hadeeth narrated by Abi Hurairah in *Sahih Bukhari*, no. 441.

[19] ابن تيمية، أحمد بن عبد الحليم بن عبد السلام. 2009. "درء تعارض العقل والنقل." دار الكتب العلمية، بيروت، لبنان

(Title in English: Averting the Conflict between Reason and Revelation).

[20] Author's translation.

[21] Surat Al-Baqarah, verse 8.

[22] Surat An-Nissaa', verse 142.

[23] Surat An-Naml, verse 14.

[24] Surat Al-Hujuraat, verse 14.

[25] Surat Al-Anfaal, verses 2-4.

[26] Surat Quraish, verse 4.

[27] Surah Youssef, verse 17.

[28] Hadeeth narrated by Ali Bin Abi Talib in *Al-Bayhaqee*, (also by Ibn Omar and Ibn Abbas).

[29] الموسوعة العقدية dorar.net/enc/aqadia/3162, accessed May 22, 2025.

[30] الشافعي، محمد بن ادريس. 1990. " كتاب الأم." دار المعارف، بيروت، لبنان

(Title in English: The Mother Book).

[31] ابن العثيمين، محمد بن صالح. 2014. " مجموع فتاوى ورسائل العثيمين." دار الثريا للنشر، الرياض، المملكة العربية السعودية

(Title in English: The Collection of Fatwas and the Messages of Ibn Uthaimeen).

[32] Surat Al-Naml, verse 14.

[33] Surat Al-Hijr, verse 99.

[34] Hadeeth narrated by Abi Hurairah in *Sahih Muslim*, no. 27.

[35] ابن قيّم الجوزيّة. محمد بن أبي بكر بن أيوب. 1997. "تهذيب مدارج السالكين." دار مؤسسة الرسالة، بيروت، لبنان

(Title in English: The Steps of the Followers).

[36] Surat Aal-Imran, verse 19.

[37] Hadeeth narrated by Abi Hurairah, in *Sahih Muslim*, no. 35, and *Sahih Bukhari*, no. 9. The quoted text is from *Sahih Muslim*.

[38] Surat Al-Hujuraat, verse 14.

[39] Surat Loqman, verse 34.

[40] Surat Al-Israa', verse 85.

[41] الحميري، عبد الملك ابن هشام. 1990. "السيرة النبوية لإبن هشام." دار الجيل، بيروت، لبنان

(Title in English: The Prophetic Biography of Ibn Hisham)

[42] Surat Yunus, verse 22.

[43] Hadeeth narrated by Abi Hurairah in *Sahih Bukhari*, no. 441.

[44] Surat Al-Baqarah, verse 170.

[45] "Doubt," Dictionary.com, Random House, accessed May 21, 2025.

[46] "Doubt," The Free Dictionary by Farlex, accessed May 18, 2025.

[47] "Doubt," *Cambridge Dictionary*, Cambridge University Press, accessed May 20, 2025.

[48] Al-Mehri, "Doubt and Certainty," *The Quran Project*, accessed May 21, 2025.

[49] Surah An-Nissaa', verse 157.

[50] Surat An-Naml, verse 66.

[51] Surat Yunus, verse 94.

[52] Surat Yunus, verse 104.

[53] Surat Ibrahim, verse 10.

[54] Surat Al-A`raaf, verses 200, 201.

[55] Surat Fussilat, verse 36.

[56] Hadeeth narrated by Abi Hurairah in *Sahih Bukhari*, no. 3276.

[57] Hadeeth narrated by Abi Hurairah in *Sahih Muslim*, no. 134.

[58] Hadeeth narrated by Abi Hurairah in *Sahih Muslim*, no. 132.

[59] البنّا، حسن. 1990. "مجموعة رسائل الإمام الشهيد حسن البنّا." دار الدعوة، الإسكندرية، مصر (Title in English: The Collection of the Messages of Imam Shaheed Hassan Al-Banna).

[60] Surat Al-A`raaf, verse 201.

[61] Surat Al-A`raaf, verse 200.

[62] Surat Fussilat, verse 36.

[63] Hadeeth narrated by Abi Hurairah in *Sahih Bukhari*, No. 4968; and in *Sahih Muslim*, No. 127. The quoted text is from S*ahih Muslim*.

[64] Hugh Pope, Faith, *The Catholic Encyclopedia*, Vol. 5 (New York: Robert Appleton Company, 1909), accessed May 21, 2025.

[65] Surat Al-Nahl, verse 125.

[66] Surat Saba', verse 24.

[67] Surat Al-Baqarah, verse 258.

[68] الجسر. نديم. 1969. " قصة الايمان بين الفلسفة والعلم والقران." دار العربية. طرابلس. لبنان. (Title in English: The Story of Belief, in Lights of Philosophy, Science, and the Qur'an)

[69] Surat Yussuf, verses 56, 57.

[70] Hadeeth narrated by Abdillah Bin Mass`ood in Al-Tabarani's *Al-Kabeer* (الكبير), and authenticated by Al- Albani in his "Authentic Series" السلسلة الصحيحة, 34.

[71] Surat Al-Anbiyaa', verse 47.

[72] Surat Ya-Seen, verse 54.

[73] Surat Fussilat, verse 53.

[74] Surat Aal-Imran, verse 26.

[75] Larry Alex Taunton, "Listening to Young Atheists: Lessons for a Stronger Christianity," The Atlantic, June 6, 2013, accessed May 22, 2025.

[76] Hadeeth narrated by Abdillah Bin Omar in *Sahih Bukhari*, no. 5767.

[77] Surat Ibrahim, verse 10.

[78] Surat Fussilat, verse 53.

[79] Surat Al-An'aam, verses 102, 103.

[80] Surat Al-Room, verse 30.

[81] Hadeeth narrated by 'Iyadh bin Himar in *Sahih Muslim*, no. 5109.

[82] Surat Yunus, verse 22.

[83] الميداني، عبد الرحمن حسن حبنكة. 1994. "العقيدة الاسلامية واسسها." دار القلم، دمشق، سوريا (Title in English: The Islamic Creed and its Foundations)

[84] Margenau, H. & R. A. Varghese (Eds.). 1993. *Cosmos, Bios, Theos: Scientists Reflect on Science, God, and the Origins of the Universe, Life, and Homo sapiens*. 285 pp. Open Court Publishing Company.

[85] Ulrich J. Becker, Ph.D., Professor of Physics, Massachusetts Institute of Technology, member of the Research Council of Europe in Geneva, Switzerland.

[86] John Erik Fornaess, Ph.D., Professor of Mathematics, Princeton University.

[87] Vera Kistiakowsky, Ph.D., Professor of Physics, Massachusetts Institute of Technology.

[88] Robert A. Naumann, Ph.D., Professor of Chemistry and Physics, Princeton University.

[89] John A. Russell, Ph.D., Distinguished Professor of Astronomy, University of Southern California.

[90] Abdus Salam, Ph.D, Nobel Prize for Physics, Director, International Center for Theoretical Physics in Trieste, Italy.

[91] Arthur L. Schowlow, Ph.D., Nobel Prize for Physics, Professor of Physics, Stanford University.

[92] Wolfgang Smith, Ph.D., Professor of Mathematics, Oregon State University.

[93] Christian B. Afinsen, Ph.D., Nobel Prize for Chemistry, Professor of

Biology, John Hopkins University.

[94] Steven L. Bernasek, Ph.D., Exxon Award in Solid State Chemistry, Professor of Chemistry, Princeton University.

[95] John Eccles, Ph.D., Nobel Prize for Physiology/Medicine.

[96] Thomas C. Emmel, Ph.D., Professor of Zoology and Director of the Division of Lepidoptera Research, University of Florida, Gainesville.

[97] Surat Aal Imarn, verse 18.

[98] Surat Aal-Imran, verse 190, 191.

[99] Hadeeth narrated by `Ataa' in *Sahih Ibn Hibbaan*.

[100] الغزالي. ابو حامد محمد. 2004. "إحياء علوم الدين." الجزء الخامس، دار المعرفة، بيروت، لبنان
(Title in English: The Revival of the Religious Sciences).

[101] Surat Al-Baqarah, verse 164.

[102] Surat Al-Room, verses 20-23.

[103] Surat Al-Nahl, verses 65-67.

[104] Surat Al-Ghashiyah, verses 17-20.

[105] Surat Al-Taareq, verse 5.

[106] Surat `Abassa, verse 24.

[107] الجسر. نديم. 1969. " قصة الايمان بين الفلسفة والعلم والقران." دار العربية. طرابلس. لبنان
(Title in English: The Story of Belief, in Lights of Philosophy, Science, and the Qur'an).

[108] الغزالي، محمد، 1900 " عقيدة المسلم." دار التراث العربي. القاهره. مصر
(Title in English: The Creed of the Muslim).

[109] الميداني. عبد الرحمن حسن حبنكة. 1994. "العقيدة الاسلامية واسسها." دار القلم. دمشق.
(Title in English: The Islamic Creed and its Foundations)

[110] Ahmad, S.F. & S.S. Ahmad, 2004. *God, Islam, and the Skeptic Mind.* p. 238. Blue Nile Publishing.

[111] Surat Ta-ha, verse 50.

[112] Surat Al-Thariyaat, verse 7.

[113] Denton, M.J. 1998. *Nature's Destiny: How the Laws of Biology Reveal Purpose in the Universe,* p. 454.

[114] Denton, M.J. 1998. *Nature's Destiny: How the Laws of Biology Reveal Purpose in the Universe,* p. 454.

[115] Surat Al-Naml, verse 88.

[116] Surat Al-Insaan, verse 1.

[117] Surat Mariam, verse 67.

[118] Surat Al-Toor, verse 35.

[119] Surat Al-Rahman, verses 26, 27.

[120] Surat Al-An'aam, verse 103.

[121] Surat Al-Shura, verse 11.

[122] ابن العثيمين، محمد بن صالح. 2014. "مجموع فتاوى ورسائل العثيمين." دار الثريا للنشر، الرياض، المملكة العربية السعودية (Title in English: The Collection of Fatwas and the Messages of Ibn Uthaimeen).

[123] Surat Al-Hujuraat, verse 15.

[124] Hadeeth narrated by Abi Hurairah in *Sahih Muslim*, no. 27.

[125] Hadeeth narrated by Abi Hurairah in *Sahih Muslim*, no. 132.

[126] Surat Al-Hijr, verse 99.

[127] Kolodiejchuk, B. (Ed.) 2007. *Mother Teresa: Come Be My Light: The Private Writings of the Saint of Calcutta.* p. 416.

[128] العودة، سلمان. 2013 "عقلي المؤمن"

[129] Baird, J. 2014. "Doubt as a Sign of Faith." New York Times, accessed May 21, 2025.

[130] Merriam-Webster, "Science," Merriam-Webster.com Dictionary, accessed May 20, 2025.

[131] Samford University, Center for Science and Religion, accessed May 21, 2025.

[132] Crystal, D. 2003. *The Cambridge Encyclopedia of the English Language.* p. 506.

[133] See "Religion," Merriam-Webster.com Dictionary, accessed May 21, 2025.

[134] See "Religion," Dictionary.com, Random House, accessed May 21, 2025.

[135] See "Religion," ScienceDaily, accessed May 21, 2025.

[136] Philip A. Pecorino, "Requirements of a Definition," *Philosophy of Religion*: Textbook by Dr. Philip A. Pecorino, Queensborough Community College, CUNY, accessed May 21, 2025.

[137] See "Religion," Academic Kids Encyclopedia, accessed May 21, 2025, https://academickids.com/encyclopedia/index.php/Religion.

[138] Ontario Consultants on Religious Tolerance, "Glossary of Religious Terms: P–R," ReligiousTolerance.org, accessed May 21, 2025.

[139] البنّا، حسن. 1990. "مجموعة رسائل الإمام الشهيد حسن البنّا." دار الدعوة، الإسكندرية، مصر (Title in English: The Collection of the Messages of Imam Shaheed Hassan Al-Banna).

[140] Surat Al-Zumar, verse 9.

[141] Surat Faatir, verse 28.

[142] Surat Al-Mujadalah, verse 11.

[143] Hadeeth narrated by Abu Hurairah, in *Sahih Bukhari*, No. 5354.

[144] بعيون، سهى. 2014. "إسهام المرأة الأندلسية في النشاط العلمي في الأندلس." الدار العربية للعلوم ناشرون. بيروت. لبنان

(Title in English: The Contribution of the Andalusian Woman to the Scientific Activities in Andalusia).

[145] بعيون، سهى. 2008. "إسهام العلماء المسلمين في العلوم في الأندلس." دار المعرفة. بيروت، لبنان

(Title in English: The Contribution of the Muslim Scholars to Sciences in Andalusia).

[146] بعيون، سهى. 2014. "إسهام المرأة الأندلسية في النشاط العلمي في الأندلس." الدار العربية للعلوم ناشرون، بيروت، لبنان

(Title in English: The Contribution of the Andalusian Woman to the Scientific Activities in Andalusia).

[147] Sabra, A. I. 2003. "Ibn al-Haytham: Brief life of an Arab mathematician." Harvard Magazine, September-October, 2003.

[148] بعيون، سهى. 2008. "إسهام العلماء المسلمين في العلوم في الأندلس." دار المعرفة. بيروت، لبنان

(Title in English: The Contribution of the Muslim Scholars to Sciences in Andalusia).

[149] بعيون، سهى. 2014. "إسهام المرأة الأندلسية في النشاط العلمي في الأندلس." الدار العربية للعلوم ناشرون. بيروت. لبنان

(Title in English: The Contribution of the Andalusian Woman to the Scientific Activities in Andalusia).

[150] هونكة. زيغريد. 1993. "شمس العرب تسطع على الغرب." دار الجليل. بيروت. لبنان

(Title in English: The Sun of the Arabs Shines over the West)

[151] السكاف. اسعد نصر الله ومحمود مطرجي. 1988. "تاريخ العلوم عند العرب." دار نظير عبود. بيروت، لبنان

(Title in English: History of Sciences of the Arabs).

[152] Surat Al-A`raaf, verse 31.

[153] Hadeeth narrated by Abdillah Bin Omar in *Sahih Muslim*, no. 2003.

[154] ابن قيّم الجوزية، محمد بن أبي بكر بن أيوب. 1992. "الطبّ النّبوي." دار مكتبة الهلال، بيروت، لبنان

(Title in English: The Prophetic Medicine)

[155] Surat Al-Qiyamah, verses 37 and 39.

[156] Bucaille, M. 2003. *The Bible, the Qur'an and Science: The Holy Scriptures Examined in the light of Modern Knowledge.* p. 272.

[157] Surat Al-An'aam, verse 162.

[158] Surat Al-Nahl, verse 89.

[159] Surat Ghafir, verse 78.

[160] Surat Al-Nahl, verse 43.

[161] Grousset, V. 1991. "Enigme: Les vertiges de la science." Le Figaro Magazine, 23 Fevrier. pp. 74-77.

[162] Surat Al-Israa', verse 85.

[163] Grousset, V. 1991. "Enigme: Les vertiges de la science." Le Figaro Magazine. 23 Fevrier. pp. 74-77.

[164] Surat Aal-Imran, verse 18.

[165] الميداني، عبد الرحمن حسن حبنكة. 1992. "صِراع مع المَلاحِدة حتى العَظْم." دار القلم، دمشق، سوريا
(Title in English: A Debate with the Atheists to the Bones).

[166] Surat Aal-Imran, verse 7.

[167] البنّا، حسن. 1990. " مجموعة رسائل الإمام الشهيد حسن البنّا." دار الدعوة، الإسكندرية، مصر
(Title in English: The Collection of the Messages of Imam Shaheed Hassan Al-Banna).

[168] الميداني، عبد الرحمن حسن حبنكة. 1992. "صِراع مع المَلاحِدة حتى العَظْم." دار القلم، دمشق، سوريا
(Title in English: A Debate with the Atheists to the Bones)

[169] Asad, M. 2003. *The Message of the Qur'an*, p. 1200.

[170] Ortoli, S. 1987. "L'évolution contestée." Science et Vie. No. 834: pp. 42-54, 162.

[171] Bowlby, J. 1992. *Charles Darwin: A New Life*. p. 528.

[172] Adler, J. 2005. Evolution of a Scientist. Newsweek. Dec. 12; Bowlby, J. 1992. *Charles Darwin: A New Life*. p. 528; Denton, M. 1986. *Evolution: A Theory in Crisis*. p. 368; Schatzman, M. 1990. "The Origin of Darwin's Despair." New Scientist, 1729.

[173] Lyell, C. 1998. *Principles of Geology: Being an Attempt to Explain the Former Changes of the Earth's Surface by Reference to Causes now in Operation*. p. 528.

[174] Denton, M. 1986. *Evolution: A Theory in Crisis,* p. 368.

[175] Wells, J. 2002. *Icons of Evolution*, p. 338.

[176] Giuseppe Sermonti (Geneticist, Professor at the University of Perousse, Director of the "Biologie Forum") in: Staune, J. 1991. "L'Evolution condamne Darwin." Le Figaro Magazine, no. 587. pp. 74-84.

[177] Denton, M. 1986. *Evolution: A Theory in Crisis.* p. 368.

[178] Jean Drost (Zoologist, Previous Director of the Museum of Natural History, Member of the French National Academy of Science) in: Staune, J. 1991. "L'Evolution condamne Darwin." Le Figaro Magazine. No. 587. pp. 74-84.

[179] Staune, J. 1991. "L'Evolution condamne Darwin." Le Figaro Magazine. No. 587. pp. 74-84.

[180] Charles Doolittle Walcott (1850 – 1927) was an American invertebrate paleontologist.

[181] Ortoli, S. 1987. "L'évolution contestée." Science et Vie. No. 834: pp. 42-54, 162.

[182] Hitching, F. 1982. "Was Darwin wrong?" Life. April 1982. V. 5, No. 4: 48-52.

[183] Denton, M. 1986. *Evolution: A Theory in Crisis,* p. 368.

[184] Sample, I. 2013. "Skull of Homo erectus Throws Story of Human Evolution in Disarray." The Guardian, October 17.

[185] Denton, M. 1986. *Evolution: A Theory in Crisis.* p. 368.

[186] See De Beer, G. 1971. *Homology, an unresolved problem.*

[187] Surat Al-Kahf, verse 51.

[188] Surat Saad, verse 71.

[189] Surat Al-Saaffaat, verse 11.

[190] Surat Al-Hijr, verse 26.

[191] Surat Al-Rahman, verse 14.

[192] Surat Saad, verse 75.

[193] Surat Aal-Imran, verse 59.

[194] Hadeeth narrated by Anas Bin Malik in *Sahih Muslim*, no. 2611.

[195] Surat Al-Hijr, verse 29.

[196] Surat Al-A'raaf, verse 27.

[197] Adler, J. 2005. Evolution of a Scientist. Newsweek. Dec. 12.

[198] Hart, M. 2000. *The 100: A Ranking of the Most Influential Persons in History,* p. 556.

[199] Jones, M. 2008. "Who Was More Important: Lincoln or Darwin?" Newsweek. July 7.

[200] Denton, M. 1986. *Evolution: A Theory in Crisis.* p. 368.

[201] Staune, J. 1991. "L'Evolution condamne Darwin," Le Figaro Magazine. No. 587. pp. 74-84.

[202] Denton, M. 1986. *Evolution: A Theory in Crisis*, p. 368.

[203] Surat Al-An'aam, verse 162.

[204] Surat Al-Nazi'aat, verse 56.

[205] Surat Aal-Imran, verse 190, 191.

[206] Surat Al-'Alaq, verse 1.

[207] الجسر، نديم. 1969. "قصة الايمان بين الفلسفة والعلم والقران." دار العربية، طرابلس، لبنان
(Title in English: The Story of Belief, in Lights of Philosophy, Science, and the Qur'an).

[208] الجسر، نديم. 1969. "قصة الايمان بين الفلسفة والعلم والقران." دار العربية، طرابلس، لبنان
(Title in English: The Story of Belief, in Lights of Philosophy, Science, and the Qur'an).

[209] Surat Al-'Alaq, verse 1.

[210] Surat Ya-Seen, verses 77-79.

[211] ابن تيمية، أحمد بن عبد الحليم بن عبد السلام. 2009. "درء تعارض العقل والنقل." دار الكتب العلمية، بيروت، لبنان
(Title in English: Averting the Conflict between Reason and Revelation).

[212] الغزالي، محمد. 1980. "عقيدة المسلم." دار التراث العربي، القاهرة، مصر
(Title in English: The Creed of the Muslim).

[213] Surat Al-Nahl, verse 53.

[214] Surat Al-Mu'minoon, verses 12-14.

[215] Surat Al-Naml, verse 62.

[216] Surat Al-Ra'd, verse 28.

[217] Surat Yussuf, verse 33.

[218] Surat Al-Insaan, verse 9.

[219] Surat Al-Ma'edah, verse 8.

[220] Surat Al-Shu'araa', verses 78—82.

[221] Hadeeth narrated by Abdullah Bin Omar in *Al-Tirmithee*, No. 2165.

[222] Surat Saba', verse 39.

[223] Surat Al-Hadeed, verse 18.

[224] Surat Al-Ankaboot, verse 2.

ABOUT CRESCENT BOOKS

Crescent Books is committed to publishing works that challenge the conventional and celebrate the diverse voices that enrich our understanding of faith, culture, and history. As a small, passionate team of book lovers, we guide authors through the publishing process with editorial freedom and genuine partnership. We serve our communities by producing books that inspire, provoke thought, and entertain, creating transformative reading experiences that connect with a broad audience and open doors to new perspectives on our ever-evolving world.

OTHER TITLES BY CRESCENT BOOKS